精通现代C++

11/14/17/20

[澳]杭小树◎编著

电子工业出版社
Publishing House of Electronics Industry
北京·BEIJING

内 容 简 介

本书旨在帮助读者深入理解现代 C++编程，全书共 10 章。第 1 章和第 2 章全面系统地介绍 C++11/14/17/20 的语言新增功能及其应用；第 3 章介绍面向对象的技术；第 4 章至第 6 章系统地介绍标准模板库（STL）的主要组成，如模板、容器、算法函数等；第 7 章介绍智能指针与内存管理；第 8 章和第 9 章介绍并发编程和并行算法，以及 CUDA 并行计算平台的入门编程知识；第 10 章介绍设计模式。

本书是关于现代 C++的高级读物，目的是提高高年级本科生、研究生、C++程序员对现代 C++的认识和编程水平。

未经许可，不得以任何方式复制或抄袭本书之部分或全部内容。
版权所有，侵权必究。
版权贸易合同登记号　图字：01-2025-3391

图书在版编目（CIP）数据

精通现代 C++11/14/17/20 ／（澳）杭小树编著.
北京：电子工业出版社，2025. 8. -- ISBN 978-7-121-50860-8

Ⅰ．TP312.8
中国国家版本馆 CIP 数据核字第 2025GP9793 号

责任编辑：田宏峰
印　　刷：三河市华成印务有限公司
装　　订：三河市华成印务有限公司
出版发行：电子工业出版社
　　　　　北京市海淀区万寿路 173 信箱　邮编 100036
开　　本：787×1 092　1/16　印张：20　字数：551 千字
版　　次：2025 年 8 月第 1 版
印　　次：2025 年 8 月第 1 次印刷
定　　价：89.00 元

凡所购买电子工业出版社图书有缺损问题，请向购买书店调换。若书店售缺，请与本社发行部联系，联系及邮购电话：（010）88254888，88258888。
质量投诉请发邮件至 zlts@phei.com.cn，盗版侵权举报请发邮件至 dbqq@phei.com.cn。
本书咨询联系方式：tianhf@phei.com.cn。

作者简介

杭小树（Hanson Hang），2002年在中国科学技术大学获模式识别与智能系统专业博士学位，2006年在澳大利亚迪肯大学（Deakin University）获计算机科学与技术专业哲学博士，2019年通过高层次人才引进回国，先后在浙江、江苏两省从事智能机器人方面的研发工作。现为江苏省产业技术研究院研究员、江苏集萃智能制造技术研究所有限公司总经理，获南京市B类人才和江苏省创新人才等荣誉。

作为中国最早一批C++程序员，早在20世纪90年代就带领团队（团队成员主要是中国科学技术大学少年班学生）为美国华尔街开发Windows平台下的商业软件。2002年出国后，先在著名的UML工具开发商Sparx Systems从事软件建模仿真与软件自动化方面的研发工作[Sparx Systems的旗舰产品Enterprise Architect（EA）深受中国用户欢迎]，后来加入世界500强ITW公司从事分布式系统的开发。

在国外期间，曾独立开发了两款大型矢量绘图软件：VDraw（最后版本5.0）和Impress（最后版本5.0）。这两款软件在国际著名软件下载网站CNET和Softpedia有数十万的下载量。

2018年起转向机器人的研发，2019年回国后带领团队在移动机器人、机械臂、复合机器人等方面从事新产品研发工作。在机器人的软件架构设计、ROS/ROS2、SLAM技术、路径规划、机器人导航与控制、人形机器人等方面有比较深入的研究，并做了大量的开发工作。近两年对人工智能新技术，如大语言模型、Transformer、强化学习、视觉语言导航、Agent等进行追踪学习和研究。

在30年的学术与技术生涯中，参与过多项国家重点自然科学基金项目和国家863项目的研发工作，在国内外发表论文30余篇，获10多项专利。

序　言

丹尼斯·里奇（Dennis Ritchie）于 1972 年在贝尔实验室开发出的 C 语言，为现代编程语言奠定了基础。C 语言以其高效、简洁和灵活性广受欢迎。本贾尼·斯特劳斯特卢普（Bjarne Stroustrup）在 1983 年开发出了 C++语言，最初被称为"C with Classes"。C++语言的核心理念是将 C 语言的高效性与面向对象的编程（OOP）结合起来，以支持更复杂的程序设计。

几十年来，C++语言一直是面向对象程序设计领域的重要支柱。从最初的扩展 C 语言的功能，到全面提供面向对象的编程支持，使得 C++语言在全球范围内获得了开发者的青睐。随着时间的推移，C++语言不断演化，从最早的 C++98 标准到后来的 C++03、C++11、C++14、C++17、C++20 和最新的 C++23。每一次的标准更新都带来了语言特性的重大革新。C++11 及以后的版本被称为现代 C++。

C++语言的应用范围广泛，涵盖了从嵌入式系统到机器人开发等多个领域。在嵌入式开发中，C++语言凭借其高效的执行性能和对硬件的精细控制，成为许多实时系统和微控制器应用的首选语言。在机器人开发中，C++语言的标准模板库和第三方库［如机器人操作系统（ROS）］提供了强大的工具和框架，支持复杂的运动控制、传感器集成和计算机视觉任务。

近年来，C++语言在编程语言排行榜上的排名持续上升，这一趋势反映了其在现代软件开发中的重要性。根据 TIOBE 编程语言排行榜和 Redmonk 的报告，C++语言已稳居全球编程语言的前列。其强大的性能、丰富的标准模板库和对底层操作的精细控制，使得 C++语言在各种复杂系统开发中不可或缺。尤其是在系统级编程、游戏开发、金融服务和高性能计算等领域，C++语言的地位更加突出。

本书旨在为广大开发者提供一部全面而深入的指南，帮助他们理解并运用现代 C++中的精髓。C++11 是 C++语言的一个重大转折点，引入了许多重要的特性，如智能指针、lambda 表达式、线程支持（并发编程）和基于范围的 for 循环。C++11 提高了代码的安全性、可读性和性能，使得开发变得更加高效。C++14 进一步优化了 C++11，增强了语言的易用性和稳定性。C++17 和 C++20 则引入了诸如结构化绑定、并行算法、概念等新特性，使得 C++编程变得更加高效和现代化。

本书作为介绍 C++编程的高级读物，除了介绍现代 C++的功能，还系统地介绍了标准模板库（STL）的模板、容器、算法函数、并行算法，以及并行计算平台 CUDA 和 OpenCL，并在最后一章以较大的篇幅详细地介绍了设计模式，为有经验的 C++开发人员提供软件架构设计方面的指导。

无论你是经验丰富的 C++程序员，还是刚刚入职的初级程序员，本书都将带你深入探索 C++语言的演进过程，通过实际示例和详尽的解释，帮助你掌握最新的语言特性和最佳实践。本书重点关注如何将现代 C++的新特性融入现有的代码库，以便最大限度地发挥现代 C++的优势，并确保代码的兼容性和可维护性。

感谢你阅读本书，希望本书能成为你掌握现代 C++的有力工具，助你在编程的道路上不断进步，达到新的高度。

由于时间仓促，书中难免出现错误，望读者批评指正。

目 录

第 1 章	语言新增功能	1
1.1	C++11 新增的语言功能	1
1.2	C++14 新增的语言功能	18
1.3	C++17 新增的语言功能	20
1.4	C++20 新增的语言功能	24
1.5	本章小结	30

第 2 章	lambda 表达式	33
2.1	C++11 中的 lambda 表达式	33
2.2	C++14 对 lambda 表达式的扩展	38
2.2.1	lambda 捕获初始化	38
2.2.2	泛型 lambda 表达式	39
2.2.3	默认参数	40
2.3	C++17 对 lambda 表达式的扩展	42
2.3.1	常量 lambda 表达式	42
2.3.2	按值捕获 this 指针	43
2.4	C++20 对 lambda 表达式的扩展	44
2.4.1	lambda 模板语法	44
2.4.2	lambda 参数包捕获	44
2.5	本章小结	45

第 3 章	面向对象的技术	47
3.1	构造函数与析构函数	47
3.1.1	转换构造函数和显式构造函数	48
3.1.2	拷贝构造函数	49
3.1.3	拷贝赋值运算符	50
3.1.4	移动构造函数	51
3.1.5	移动赋值运算符	52
3.1.6	委派构造函数	53
3.1.7	继承构造函数	53
3.1.8	显式默认函数和显式删除函数	54
3.1.9	私有构造函数	55
3.2	虚函数与多态性	56
3.2.1	虚函数表和虚函数指针	58
3.2.2	显式重写和终止重写	58
3.2.3	常量表达式虚函数	59
3.3	重载	61
3.3.1	重载赋值运算符	61
3.3.2	函数调用运算符的重载	62
3.4	继承	63
3.4.1	多重继承	63
3.4.2	多级继承	64
3.4.3	虚继承	65
3.5	其他杂项	66
3.5.1	左值引用和右值引用	66
3.5.2	移动语义	67
3.5.3	引用限定符	69
3.6	本章小结	71

第 4 章	模板	73
4.1	函数模板	74
4.1.1	函数模板实例化	75
4.1.2	函数模板实参推导	76
4.1.3	显式函数模板实参	76
4.1.4	函数模板实参替换	77
4.1.5	函数模板实参重写	77
4.1.6	参数包	79
4.1.7	折叠表达式	80

- 4.2 类模板 ··· 81
  - 4.2.1 显式类模板实例化 ············ 82
  - 4.2.2 隐式类模板实例化 ············ 82
  - 4.2.3 类模板形参和类模板实参 ································ 84
  - 4.2.4 类模板实参推导 ··············· 85
  - 4.2.5 非类型模板参数中的类类型 ································ 85
  - 4.2.6 用 auto 声明非类型模板参数 ································ 85
- 4.3 类型别名和别名模板 ··············· 86
- 4.4 变量模板 ··············· 88
- 4.5 概念与约束 ··············· 89
  - 4.5.1 概念 ··············· 89
  - 4.5.2 约束 ··············· 91
  - 4.5.3 约束表达式 ··············· 93
- 4.6 本章小结 ··············· 96

第 5 章 STL 容器 ··············· 99

- 5.1 顺序容器 ··············· 99
  - 5.1.1 std::array ··············· 100
  - 5.1.2 std::vector ··············· 101
  - 5.1.3 std::list ··············· 103
  - 5.1.4 std::forward_list ··············· 105
  - 5.1.5 std::deque ··············· 107
- 5.2 关联容器 ··············· 108
  - 5.2.1 std::set ··············· 109
  - 5.2.2 std::multiset ··············· 110
  - 5.2.3 std::map ··············· 111
  - 5.2.4 std::multimap ··············· 113
- 5.3 无序关联容器 ··············· 113
  - 5.3.1 std::unordered_set ··············· 114
  - 5.3.2 std::unordered_map ··············· 115
  - 5.3.3 std::unordered_multiset ··············· 117
  - 5.3.4 std::unordered_multisetmap ··············· 119
- 5.4 容器适配器 ··············· 120
  - 5.4.1 std::stack ··············· 120
  - 5.4.2 std::queue ··············· 121
  - 5.4.3 std::priority_queue ··············· 122
- 5.5 分配器与迭代器 ··············· 126
  - 5.5.1 std::allocator ··············· 126
  - 5.5.2 迭代器 ··············· 126
  - 5.5.3 迭代器失效 ··············· 127
- 5.6 本章小结 ··············· 127

第 6 章 STL 函数 ··············· 129

- 6.1 算法函数 ··············· 129
  - 6.1.1 排序算法 ··············· 129
  - 6.1.2 搜索算法 ··············· 130
  - 6.1.3 非更改顺序算法 ··············· 131
  - 6.1.4 更改顺序算法 ··············· 135
  - 6.1.5 分割算法 ··············· 140
  - 6.1.6 合并算法 ··············· 141
  - 6.1.7 堆算法 ··············· 142
  - 6.1.8 最大最小值算法 ··············· 144
  - 6.1.9 数值算法 ··············· 146
- 6.2 函数对象 ··············· 148
  - 6.2.1 std::greater 和 std::less ··············· 149
  - 6.2.2 std::reference_wrapper ··············· 149
  - 6.2.3 std::ref 和 std::cref ··············· 150
- 6.3 Utility 函数 ··············· 151
  - 6.3.1 std::move ··············· 151
  - 6.3.2 std::forward ··············· 152
  - 6.3.3 std::swap ··············· 153
  - 6.3.4 std::make_pair ··············· 153
- 6.4 回调函数 ··············· 154
  - 6.4.1 回调函数的基本概念 ··············· 154
  - 6.4.2 使用普通函数实现回调函数 ··············· 154
  - 6.4.3 使用函数指针实现回调函数 ··············· 155
  - 6.4.4 使用函数对象实现回调函数 ··············· 155
  - 6.4.5 将 lambda 表达式传入回调函数 ··············· 157

6.4.6 使用 std::bind 实现回调函数 ······ 158
6.4.7 使用 std::function 实现回调函数 ······ 159
6.5 本章小结 ······ 162

第 7 章 智能指针与内存管理 ······ 165

7.1 堆栈和内存分配 ······ 165
7.2 指针与内存泄漏 ······ 166
7.3 分段错误 ······ 168
7.4 智能指针 ······ 169
    7.4.1 std::unique_ptr ······ 170
    7.4.2 std::shared_ptr ······ 177
    7.4.3 std::weak_ptr ······ 183
7.5 本章小结 ······ 185

第 8 章 并发与多线程 ······ 187

8.1 并发与并行 ······ 187
8.2 创建线程 ······ 188
8.3 线程同步与互斥 ······ 192
    8.3.1 std::mutex ······ 192
    8.3.2 std::condition_variable ······ 193
    8.3.3 std::lock_guard 和 std::unique_lock ······ 194
    8.3.4 std::atomic ······ 199
8.4 线程死锁 ······ 200
    8.4.1 std::lock ······ 200
    8.4.2 std::scoped_lock ······ 203
8.5 STL 中的 <future> ······ 204
    8.5.1 std::async ······ 204
    8.5.2 std::future ······ 206
    8.5.3 std::promise ······ 207
    8.5.4 std::packaged_task ······ 208
8.6 线程池 ······ 210
8.7 本章小结 ······ 213

第 9 章 并行算法与并行计算 ······ 215

9.1 STL 并行算法 ······ 215
    9.1.1 std::execution::seq ······ 215
    9.1.2 std::execution::par ······ 215
    9.1.3 std::execution::par_unseq ······ 216
    9.1.4 std::execution::unseq ······ 216
9.2 常用的并行算法 ······ 217
    9.2.1 std::sort ······ 217
    9.2.2 std::transform ······ 218
    9.2.3 std::find、std::find_if 和 std::find_if_not ······ 219
    9.2.4 std::search ······ 220
9.3 C++17 中新增的并行算法 ······ 221
    9.3.1 std::for_each 和 std::for_each_n ······ 222
    9.3.2 std::reduce 和 std::transform_reduce ······ 223
    9.3.3 std::exclusive_scan 和 std::inclusive_scan ······ 227
    9.3.4 std::transform_exclusive_scan 和 std::transform_inclusive_scan ······ 229
9.4 CUDA 并行计算编程 ······ 230
9.5 OpenCL 编程 ······ 237
9.6 本章小结 ······ 244

第 10 章 设计模式 ······ 247

10.1 设计模式概念 ······ 247
10.2 创建设计模式 ······ 248
    10.2.1 工厂方法 ······ 249
    10.2.2 抽象工厂方法 ······ 251
    10.2.3 构建器方法 ······ 253
    10.2.4 原型方法 ······ 257
    10.2.5 单例方法 ······ 259
10.3 结构设计模式 ······ 261
    10.3.1 适配器方法 ······ 261
    10.3.2 桥接方法 ······ 263
    10.3.3 组合方法 ······ 265
    10.3.4 装饰器方法 ······ 269
    10.3.5 门面方法 ······ 271
    10.3.6 代理方法 ······ 274

10.3.7　蝇量级方法 …………… 276
10.4　行为设计模式 …………… 279
　　　10.4.1　责任链方法 …………… 279
　　　10.4.2　迭代器方法 …………… 282
　　　10.4.3　中介器方法 …………… 285
　　　10.4.4　备忘录方法 …………… 288
　　　10.4.5　观察者方法 …………… 291
　　　10.4.6　状态方法 ……………… 295

　　　10.4.7　策略方法 ……………… 297
　　　10.4.8　模板方法 ……………… 299
　　　10.4.9　命令方法 ……………… 302
　　　10.4.10　访客方法 …………… 306
10.5　本章小结 ………………… 310

参考文献 …………………………… 311

第 1 章
语言新增功能

C++标准委员会成立后每隔三年就会发布一个新版本。C++11 及以后的 C++语言被称为现代 C++（Modern C++）。C++11 是 C++语言的第二个主要版本，也是自 C++98 以来最重要的更新。C++11 对编程实践进行了标准化，并改进了 C++语言中的抽象。每一个新版本的新增功能都分为语言新增功能和标准模板库（Standard Template Library，STL）新增功能。本章重点介绍 C++11/14/17/20 等版本的语言新增功能，STL 新增功能在后面的章节专门介绍。本章的内容安排顺序与 C++版本的顺序一致，以便读者了解每个版本的新增功能。

1.1 C++11 新增的语言功能

在 C++11 的语言新增功能中，有一些比较重要的功能，如 lambda 表达式将在第 2 章介绍；转换构造函数（converting constructor）、移动特殊成员函数（move special member function）、引用限定符（ref-qualifier）将在第 3 章介绍；变量模板（variable template）将在第 4 章介绍；移动（move）和转发（forward）将在第 6 章介绍。本节对 C++11 的语言新增功能进行逐一介绍，并给出示例代码。

1. 初始化列表（initializer lists）

在使用初始化列表时，需要区分构造函数中的成员变量初始化列表和 std::initializer_list。下面先介绍构造函数中的成员变量初始化列表。

在使用参数构造函数时，可使用初始化列表进行成员变量的初始化，例如：

```cpp
class Point{
private:
    int x;
    int y;
public:
    Point(int i = 0, int j = 0): x(i), y(j) {}
    int getX() const { return x; }
    int getY() const { return y; }
};
```

常量成员变量必须使用初始化列表进行初始化，例如：

```cpp
class TestA {
    const int t;
public:
```

```
        TestA(int t0):t(t0) {}          //必须使用初始化列表对常量成员变量进行初始化
        int getT() { return t; }
};
```

引用成员变量时必须使用初始化列表进行初始化，例如：

```
class TestB {
    int &t;
public:
    TestB(int &t):t(t) {}              //必须使用初始化列表对引用的成员变量进行初始化
    int getT() { return t; }
};
```

参数化的基类构造必须使用初始化列表进行初始化，例如：

```
class A {
    int i;
public:
    A(int );
};
A::A(int k){
    i = k;
}
class B: public A{
public:
    B(int );
};
B::B(int x):A(x) {                     //必须使用初始化列表对参数化的基类构造进行初始化
}
```

std::initializer_list 是一个类模板，使用"支撑列表"语法创建轻量级的类似数组的元素容器。例如，通过{1, 2, 3}可创建一个类型为 std::initializer_list <int>的整数序列。std::initializer_list<T>提供了一种访问任何 const T 的数组。注意：在使用 std::initializer_list 时需要加上#include<initializer_list>。

```
#include<initializer_list>
template<class T>
struct S{
    std::vector<T> v;
    S(std::initializer_list<T> t) : v(t){}
    void append(std::initializer_list<T> t){
        v.insert(v.end(), t.begin(), t.end());
    }
};

int sum(const std::initializer_list<int>& list){
    int total = 0;
    for (auto& e : list) {
        total += e;
    }
    return total;
```

```
auto list = {1, 2, 3};        //list 相当于 std::initializer_list<int>
sum(list);                    //返回值为 6
sum({1, 2, 3});               //{1, 2, 3}的类型为 const std::initializer_list<int>
sum({});                      //返回值为 0
```

2. 静态断言（static_assert）

在编译程序时编译器会检查断言，因此需要声明静态断言。如果声明静态断言失败，则会导致程序格式不正确，并可能生成诊断错误消息。静态断言有以下三种形式：

（1）具有固定错误消息的静态断言。
（2）没有错误消息的静态断言。
（3）带有用户生成错误消息的静态断言。

static_assert 声明可以出现在命名空间、块作用域（作为块声明）以及类中。在 C++11 中，静态断言的语法格式如下：

```
static_assert (bool-constexpr, unevaluated-string)
```

这里，bool-constexpr 是返回布尔值的表达式；unevaluated-string 是未计算的字符串。未计算的字符串是指字符串中不带有任何编码前缀的字符串文字。如果 bool-constexpr 返回布尔值 false，则显示编译错误信息 unevaluated-string。

```
constexpr int x = 0;
constexpr int y = 1;
static_assert(x == y, "x != y");

template <class T, int Size>
class Vector {
    /*编译时检查 Vector 的大小（Size）是否大于 3，如果 Size < 4，则声明静态断言失败，显示编译错误"Vector size is too small!" */
    static_assert(Size > 3, "Vector size is too small!");

private:
    T m_values[Size];
};

template<class T>
void swap(T& a, T& b) noexcept{
    static_assert(std::is_copy_constructible_v <T>, "Swap requires copying");
    auto c = b;
    b = a;
    a = c;
}
```

3. 关键字 auto

C++11 引入的关键字 auto 使开发人员能够在不显式指定数据类型的情况下声明变量，编译器可以从变量的初始值或初始化表达式中推断变量的类型。该功能非常方便，不仅简化了代码，还增加了代码的可读性，尤其是当数据类型比较复杂或从上下文中显而易见时。

初始化 auto 类型变量的代码如下：

```cpp
auto a = 1;                      //int
auto b = 3.14;                   //double
auto& c = b;                     //double&
const auto d = 1;                //const int
auto e {0};                      //std::initializer_list<int>
auto&& f = 1;                    //int&&
auto&& g = b;                    //double&&
auto h = new auto(123);          //int*const
```

关键字 auto 的用途广泛,可用于任何类型,包括函数、迭代器、lambda 函数等。例如:

```cpp
std::vector<int> v{3, 6, 8, 4, 5, 9};
for(auto iter = v.begin(); iter != v.end(); it++)
cout << *iter;
auto fun_sum = [](int a, int b){
    return a + b;
};
```

使用 auto 作为函数返回值的类型时,请使用尾随返回类型,例如:

```cpp
auto add(int x, int y) -> int{
    return x + y;
}
auto a = add(1, 2);
```

4. 类型说明符 decltype

decltype 是一个类型说明符,可返回传递给它的表达式的声明类型。例如:

```cpp
int a = 1;
decltype(a) b = a;
const int& c = a;
decltype(c) d = a;               //decltype(c) is const
decltype(123) e = 123;           //decltype(123) is int
int&& f = 1;                     //f is declared as type int&&
decltype(f) g = 1;               //decltype(f) is int&&
decltype((a)) h = g;             //decltype((a)) is int&
```

decltype 还可以与 auto 一起推断函数返回类型。在 C++11 中,必须显式地指定返回类型或者使用 decltype。例如:

```cpp
template <typename X, typename Y>
auto add(X x, Y y) -> decltype(x + y){
    return x + y;
}
add(1, 2);
```

5. 类型别名(type alias)

类型别名是指引用先前定义的类型的名称(类似于 typedef),别名模板(alias template)是引用一系列类型的名称。语法格式如下:

```cpp
using identifier attr(optional) = type-id;
template <template-parameter-list>
```

```
using identifier attr(optional) = type-id;
```

其中的 attr 是可选项。类型别名示例如下：

```
using ULINT = unsigned long int;
ULINT a = 123456789;
ULINT b = 987654321;
ULINT c = 74;

template <typename T>
using Vec = std::vector<T>;
Vec<int> v;                        //等价于 std::vector<int> v;

using func = void (*) (int, int);  //等价于 typedef void (*func)(int, int);

//使用别名模板的隐藏参数
template<class CharT>
using mystring = std::basic_string<CharT, std::char_traits<CharT>>;
mystring<char> str;

//将别名类型引入类成员变量数据类型
template<typename T>
struct Container {
    using value_type = T;
};
```

6. 空指针（nullptr）

C++11 之前的版本使用 NULL 表示空指针，但这种方法存在明显的缺陷，编译器在实现时常常将其定义为 0，并导致重载混乱。例如下面的代码：

```
void f(char*);
void f(int);
```

当调用 f(NULL)时编译器会匹配到 f(int)函数，这显然会让人感到迷惑。C++11 引入了关键字 nullptr（类型为 nullptr_t），可区分空指针与 0，且 nullptr 能够隐式地转换为任何类型的指针或成员指针的类型。

【例 1-1】

```
#include<iostream>
using namespace std;
int main(){
    int *ptr = nullptr;
    if (ptr){
        cout << "true";
    }
    else {
        cout << "false";
    }
    return 0;
}
```

【例1-2】

```cpp
#include <iostream>
using namespace std;
int main(){
    nullptr_t np1, np2;
    if (np1 >= np2)
        cout << "can compare" << endl;
    else
        cout << "can not compare" << endl;
    char *x = np1;
    if (x == nullptr)
        cout << "x is null" << endl;
    else
        cout << "x is not null" << endl;
    return 0;
}
```

7．强类型枚举（strongly-typed enum）

C++11新增了强类型枚举（也称为有作用域枚举），其用法如下：

```cpp
enum struct|class name {enumerator = constant-expression, enumerator = constant-expression , ...};
enum struct|class name:type {enumerator = constant-expression, enumerator = constant-expression , ...};
enum struct|class name ;
enum struct|class name : type ;
enum class fruit { orange, apple };
enum class Color { red, green = 20, blue };
Color r = Color::blue;
switch(r){
    case Color::red   : std::cout << "red\n";   break;
    case Color::green: std::cout << "green\n"; break;
    case Color::blue : std::cout << "blue\n";   break;
}

enum class Altitude: char{
    high = 'h',
    low = 'l',              //在CWG 518后允许尾随逗号
};
Altitude a = Altitude::high;
switch(a){
    case Altitude::high :
    break;
    case Altitude::low:
    break;
}
```

8．属性（attribute）

属性是现代C++的关键功能之一，它允许程序员向编译器指定附加信息以强制执行约束（条件）、优化某些代码段或执行某些特定的代码。简单来说，属性充当了编译器的注释，它提供代码的附加信息，以便进行优化并对其强制执行某些条件。属性是C++语言的最佳功能之一，并且随

着 C++语言的版本更新而不断发展，其语法格式：

```
[[ attribute-list ]]
```

属性包括以下几种：

（1）简单属性，如[[noreturn]]。
（2）带命名空间的属性，如[[gnu::unused]]。
（3）带参数的属性，如[[deprecated（"because"）]]。
（4）带命名空间和参数的属性，如[[CC::opt(1), CC::debug]]。

以下属性是由 C++标准定义的，这些属性不能包含语法错误，必须应用于正确的目标，并且参数中的实体必须符合单一定义规则（ODR）。这些属性及其对应的 C++版本如下：。

- [[noreturn]]：对应的版本是 C++11。
- [[carries_dependency]]：对应的版本是 C++11。
- [[deprecated]]：对应的版本是 C++14。
- [[deprecated("reason")]]：对应的版本是 C++14。
- [[fallthrough]]：对应的版本是 C++17。
- [[maybe_unused]]：对应的版本是 C++17。
- [[nodiscard]]：对应的版本是 C++17。
- [[nodiscard("reason")]]：对应的版本是 C++20。
- [[likely]]：对应的版本是 C++20。
- [[unlikely]]：对应的版本是 C++20。
- [[no_unique_address]]：对应的版本是 C++20。
- [[assume(expression)]]：对应的版本是 C++23。

属性的作用有以下几点：

（1）对代码强制执行约束。这里的约束是指特定函数的参数必须满足条件才能执行（前提条件）。例如：

```
int f(int i)[[expects:i > 0]]
{
    //Code
}
```

（2）为了优化代码，向编译器提供额外的信息。编译器非常擅长优化代码，但与人类相比，它在某些地方仍然落后，而且会生成效率并不高的通用代码。这主要是由于编译器缺乏有关人类对同一问题的额外信息。为了在一定程度上减少这个问题，C++标准引入了一些新属性，允许向编译器而不是代码语句本身指定更多内容。例如：

```
int f(int i)
{
    switch (i) {
        case 1:
            [[fallthrough]];
            [[likely]]
        case 2 : return 1;
    }
```

```
        return -1;
}
```

（3）可以防止程序员故意在代码中添加某些警告和错误信息。

【例 1-3】

```cpp
#include <iostream>
#include <string>

int main(){
    //Set debug mode in compiler or 'R'
    [[maybe_unused]] char mg_brk = 'D';
    //编译器不会给出关于这个未使用变量的任何警告或错误信息
    return 0;
}
```

9. 关键字 constexpr

在 C++11 中，关键字 constexpr 的功能十分强大，允许在编译程序时评估函数或变量的值。该关键字可以应用于变量、函数和构造函数，使它们能够在常量表达式中使用。常量表达式是由编译器计算的表达式。constexpr 的语法如下：

```cpp
constexpr variable;
constexpr function;

constexpr int a[10] = {0, 1, 2, 3, 4, 5, 6, 7, 8, 9};
constexpr int length = sizeof(a)/sizeof(int);

constexpr int foo();
constexpr bool b1 = noexcept(foo());        //false，未定义 constexpr 函数

constexpr int foo() {return 1;}
constexpr bool b2 = noexcept(foo());        //true，foo()是常量表达式

constexpr int factorial(int n){
    return n <= 1 ? 1 : (n * factorial(n - 1));
}
```

10. 委派构造函数（delegating constructors）

委派构造函数是指构造函数能够调用同一个类的其他构造函数。委派构造函数是 C++11 中的新增功能。如果类的构造函数被重载，且必须初始化列表中实例的属性，则必须使用委派构造函数。例如：

```cpp
class TestClass {
private:
    const int myVariable;
public:
    TestClass(): TestClass(5) {}    //使用委派构造函数对 myVariable 进行初始化
    TestClass(int val) : myVariable(val) {}
    void display(){
        cout << "Value of myVariable is:" << myVariable << endl;
```

```cpp
        }
};

class A {
public:
    A() {}
    //成员初始化列表,非委派构造函数
    A( string str) : m_string{str} {}
    //错误,不可以在使用委派构造函数的同时使用初始化列表
    A( string str, double dbl) : A(str), m_double{dbl} {}
    //正确,可以使用赋值语句
    A(string str, double dbl) : A(str) {
        m_double = dbl;
    }
private:
    double m_double;
    string m_string;
};
```

11. 用户定义的文本(user-defined literals,UDL)

用户定义的文本可为各种内置类型提供文本,这些类型仅限于整数、字符、浮点数、字符串、布尔值和指针。简单来说,它们将值与单位相结合。UDL 被视为对文本运算符的调用,并且仅支持后缀形式。文字运算符的名称是 operator"" 后跟后缀。

【例 1-4】

```cpp
#include <iomanip>
#include <iostream>
using namespace std;

//kilogram(kg)
long double operator"" _kg(long double x){
    return x * 1000;
}

//gram(g)
long double operator"" _g(long double x) { return x;}

//miligram(mg)
long double operator"" _mg(long double x){
    return x / 1000;
}

int main(){
    long double weight = 3.6_kg;
    cout << weight << endl;
    cout << setprecision(8) << (weight + 2.3_mg) << endl;
    cout << (32.3_kg / 2.0_g) << endl;
    cout << (32.3_mg * 2.0_g) << endl;
```

```
        return 0;
}
```

12. 左值引用和右值引用（lvalue reference 和 rvalue reference）

在 C++中，每个表达式要么是左值，要么是右值。左值表示内存资源无法重用的对象，右值表示内存资源可以重用的对象。左值引用是绑定到左值的引用，用一个&标记。右值引用是绑定到右值的引用，用两个&&标记。例如：

```
int a = 10;                 //a 是整型左值
int& lref = a;              //左值引用
int&& rref = 20;            //右值引用
```

左值引用的用途为：
（1）左值引用可为现有对象设置别名。
（2）左值引用可实现按引用传递语义。
右值引用的用途为：
（1）右值引用可延长分配给它的临时对象的生存期。
（2）当把右值引用对象作为参数传入函数时，意味着不需要再关心它了。

【例 1-5】

```
#include <iostream>
using namespace std;

void printLValueReference(int& x){
    cout << x << endl;
}
void printRValueReference(int&& x){
    cout << x << endl;
}

int main(){
    int a {10};
    printLValueReference(a);
    printRValueReference(std::move(a));
    return 0;
}

int x = 0;                  //x 是整型左值
int& xl = x;                //xl 是左值引用
int&& xr = x;               //编译错误，因为 xr 是右值引用，而 x 是整型左值
int&& xr2 = 0;              //xr2 是右值引用，绑定到右值 0

void f(int& x) {}
void f(int&& x) {}
f(x);                       //调用 f(int&)
f(xl);                      //调用 f(int&)
f(3);                       //调用 f(int&&)
f(std::move(x));            //调用 f(int&&)
f(std::move(xr2));          //调用 f(int&& x)
```

13. 标识符 override

C++11 引入了标识符 override，当编译器遇到 override 时，会把 override 理解成同一个基类的重写版本。这将导致编译器检查基类，查看该基类是否有具有此确切签名的虚函数。如果没有，编译器将显示错误。通过 override，可以使编译器明确程序员的意图，如果 override 是与成员函数一起使用的，则编译器会确保基类中存在成员函数，并且还会限制程序以其他方式进行编译。

【例 1-6】

```cpp
#include <iostream>
using namespace std;

class Base {
public:
    Base(){}
    virtual void func() { cout << "I am in Base" << endl; }
    virtual ~Base (){ std::cout << "Base::~Base()" << endl;}
};

class Derived : public Base {
public:
    Derived(){}
    void func() override{
        cout << "I am in Derived class" << endl;
    }
    ~Derived()override {std::cout << "Derived::~Derived()" << endl;}
};
```

14. 说明符 final

C++11 允许使用说明符 final 来防止派生类覆盖基类的虚函数，或者阻止某个类被继承。例如：

```cpp
class Base{
public:
    virtual void f(){
        std::cout << "Base class default behaviour\n";
    }
};

class Derived : public Base{
public:
    void f() final{                    //void f()是最终版本，不能再被继承
        std::cout << "Derived class overridden behaviour\n";
    }
};
```

如果不希望一个类被继承，则可以使用 final，例如：

```cpp
class X final{
    //...
};
```

15. 默认（default）函数

显式默认函数声明是 C++11 标准新增的一种新的函数声明形式，它允许程序员将 "=default;" 添加到函数声明的末尾，从而将该函数声明为显式默认函数。这使得编译器为显式默认函数生成默认实现，这比手动编程的函数实现更有效。

【例 1-7】

```cpp
#include <iostream>
using namespace std;

class A{
public:
    A() = default;
    A(int x){ cout << "This is a parameterized constructor";}
};

int main(){
    A a;
    A x(1);
    return 0;
}
```

不是任何函数都可以被设置为默认函数。默认函数必须是特殊的成员函数（默认构造函数、拷贝构造函数、析构函数等）或者没有默认参数。例如，以下代码说明不能默认使用非特殊成员函数。

```cpp
class B {
public:
    int func()     = default;      //编译错误，一般函数不可以被设置为默认函数
    B(int, int) = default;         //编译错误，B(int, int)不是特殊的成员函数
    B(int x = 0)   = default;      //编译错误，B(int x=0)有默认参数
};

struct C{
    C(const C&) = default;
    //C::C()是隐式删除函数
};
```

16. 说明符 delete

C++11 标准新增了说明符 delete，在成员函数后面附加 "=delete" 可禁用该成员函数。使用 "=delete" 禁用的成员函数称为显式删除函数。说明符 delete 通常是针对隐式函数的，并不限于成员函数，例如可以使用=delete 来禁用拷贝构造函数和拷贝赋值函数。

【例 1-8】

```cpp
#include <iostream>
using namespace std;

class A {
public:
    A(int x): m(x){}
```

```
        A(const A&) = delete;
        A& operator=(const A&) = delete;
        int m;
};

int main(){
    A a1(1), a2(2);
    a1 = a2;              //错误，拷贝赋值函数已被删除
    A a3 = A(a2);         //错误，拷贝构造函数已被删除

    return 0;
}
```

17．基于范围的 for 循环

在某个范围内执行 for 循环，可在保持与传统 for 循环等价性的基础上，赋予传统 for 循环更强的可读性。其语法格式为：

attr (optional) for (init-statement (optional) range-declaration : range-expression) loop-statement;

【例 1-9】

```
#include <iostream>
int main(){
    std::vector<int> v = {0, 1, 2, 3, 4, 5};

    for (auto i : v)
    std::cout << i << ' ';
    std::cout << '\n';

    for (int i : {0, 1, 2, 3, 4, 5})
    std::cout << i << ' ';
    std::cout << '\n';

    for (const int& i : v)
    std::cout << i << ' ';
    std::cout << '\n';

    int a[] = { 0, 1, 2, 3, 4, 5 };
    for (int n : a)
    cout << n << ' ';
    std::cout << '\n';

    string str = "Hello C++";
    for (char c : str)
    cout << c << ' ';
    std::cout << '\n';

    std::map<int, int> MAP({ { 1, 1 }, { 2, 2 }, { 3, 3 } });
    for (auto m : MAP)
```

```
        cout << '{' << m.first << ", " << m.second << "}\n";
        return 0;
}
```

18. 关键字 explicit

在 C++11 中，可以通过关键字 explicit 来强调构造函数不能进行隐式转换。通常情况下，如果某个类的构造函数只有一个参数，那么可以通过隐式转换来构造该类的实例。但可以使用 explicit 来禁止隐式调用转换构造函数。

【例 1-10】

```cpp
#include <iostream>
using namespace std;

struct A{
    A(int) {}                                  //转换构造函数
    A(int, int) {}                             //转换构造函数（C++11）
    operator bool() const { return true; }
};
struct B{
    explicit B(int) {}
    explicit B(int, int) {}
    explicit operator bool() const { return true; }
};

int main(){
    A a1 = 1;                                  //正确：拷贝初始化选择 A::A(int)
    A a2(2);                                   //正确：直接初始化选择使用 A::A(int)
    A a3 {4, 5};                               //正确：直接初始化选择使用 A::A(int, int)
    A a4 = {4, 5};                             //正确：拷贝初始化选择 A::A(int, int)
    A a5 = (A)1;                               //正确：使用 static_cast 进行显式类型转化
    if (a1) { }                                //正确：调用 A::operator bool()
    bool na1 = a1;                             //正确：拷贝初始化选择 A::operator bool()
    bool na2 = static_cast<bool>(a1);          //正确：使用 static_cast 直接初始化
    //B b1 = 1;                                //错误：拷贝初始化不能使用 B::B(int)
    B b2(2);                                   //正确：直接初始化选择使用 B::B(int)
    B b3 {4, 5};                               //正确：直接初始化选择使用 B::B(int, int)
    //B b4 = {4, 5};                           //错误：拷贝初始化不能使用 B::B(int, int)
    B b5 = (B)1;                               //正确：使用 static_cast 进行显式类型转化
    if (b2) { }                                //正确：使用 B::operator bool()
    //bool nb1 = b2;                           //错误：拷贝初始化不能选择 B::operator bool()
    bool nb2 = static_cast<bool>(b2);          //正确：使用 static_cast 直接初始化
    return 0;
}
```

19. 内联命名空间（inline-namespace）

在 C++中，命名空间可以嵌套，且命名空间变量的解析是分层的。内联命名空间是在其原始命名空间的定义中使用可选关键字 inline 的命名空间，这使得嵌套内联命名空间标识符与父/封闭命名空间标识符的作用一样。内联命名空间的语法如下：

```
inline namespace namespace_name;
```

【例 1-11】

```cpp
#include <iostream>
namespace parent_ns {
    inline namespace nested_ns1 {
        namespace nested_ns2 {
            void func() {
                std::cout << "Function from Old Defintion\n";
            }
        }
    }
}

//使用新语法定义同样的命名空间
namespace new_parent_ns ::inline namespace new_nested_ns1 :: namespace new_nested_ns2 {
    void func(){
        std::cout << "Function from New Definition";
    }
}

int main(){
    old_parent::old_nested_ns2::func();
    new_parent::new_nested_ns2::func();
    return 0;
}
```

20. 非静态数据成员初始化器（non-static data member initializer）

类中的非静态数据成员变量通常有两种初始化方式：一是通过构造函数的成员变量初始化列表进行初始化；二是通过大括号{}或者等号=，在成员变量声明时设定初始值。如果成员变量没有使用初始值，则使用成员变量初始化列表。例如：

```cpp
class S {
    int n = 1;                      //使用{}或=
public:
    S() {}
    S(int arg) : n(arg) {}          //使用成员变量初始化列表
};

class A{
    int n = 7;                      //使用{}或=
    std::string s{'a', 'b', 'c'};   //使用{}
public:
    A(){}
};
```

21. 右尖括号（right angle bracket）

在 C++11 之前，连续的两个右尖括号会引起编译错误，中间必须用空格隔开。从 C++11 起，连续的两个右尖括号可以在一起而不会引起编译错误。例如：

```cpp
#include <vector>
typedef std::vector<std::vector<int> > vec1;           //正确
typedef std::vector<std::vector<bool>> vec2;           //在 C++11 之前是错误的
typedef std::vector<std::vector<bool>> vec3;           //自 C++11 起是正确的
```

22. 尾随返回类型（trailing return type）

C++11 新增了一种函数返回类型方法，即尾随返回类型。在 C++11 之前，函数的返回类型通常在函数名称之前指定。但是，在某些情况下，表达复杂的返回类型可能具有挑战性，尤其是在处理模板函数或带有 decltype 的函数时。使用尾随返回类型，可以通过关键字 auto 和尾随箭头 -> 在参数列表之后指定函数的返回类型。这允许程序员使用表达式和 decltype 根据函数的实现来推断返回类型。例如：

```cpp
auto function_name(parameter_list) -> return_type{
    //函数实现
    //...
}
```

【例 1-12】

```cpp
#include <iostream>
auto add(int a, int b) -> decltype(a + b) { return a + b; }

int main(){
    int x = 5, y = 10;
    auto result = add(x, y);
    cout << "Result: " << result << endl;
    return 0;
}
```

23. 说明符 noexcept

说明符 noexcept 用于表示函数不会引发任何异常，对编译器和程序员都很有价值。通过 noexcept 可以为编译器提供保证，即不会从该函数传播异常，从而允许编译器优化代码并生成更高效的二进制文件。此外，使用代码的其他开发人员将认识到，在需要防止异常的情况下，调用 noexcept 函数是安全的。noexcept 的语法格式如下：

（1）noexcept。
（2）noexcept(expression)。

```cpp
void f() noexcept;
void f();                   //错误：已经声明了 noexcept 同名函数
void g() noexcept(false);
void g();                   //正确，g 的两个声明都可能抛出异常

struct B{
    virtual void f() noexcept;
    virtual void g();
    virtual void h() noexcept = delete;
};
struct D: B{
    void f();               //ill-formed: D::f 可能抛出异常，B::f 不会抛出异常
```

```
        void g() noexcept;        //正确
        void h() = delete;        //正确
};
```

24. char32_t 和 char16_t

char32_t 用于表示 32 位宽字符的无符号整型数据，与 uint_least32_t 功能相同。char16_t 用于表示 16 位宽字符的无符号整型数据，与 uint_least16_t 功能相同。

```
typedef uint_least32_t char32_t;
typedef uint_least16_t char16_t;
```

【例 1-13】

```
#include <stdio.h>
#include <uchar.h>
int main(void){
    const char32_t wc[] = U"zß 水□";
    const size_t wc_sz = sizeof wc / sizeof *wc;
    printf("%zu UTF-32 code units: [ ", wc_sz);
    //输出：5 UTF-32 code units: [ 0x7a 0xdf 0x6c34 0x1f34c 0 ]
    for (size_t n = 0; n < wc_sz; ++n)
        printf("%#x ", wc[n]);
    printf("]\n");
    return 0;
}
```

【例 1-14】

```
#include <stdio.h>
#include <uchar.h>
int main(void){
    const char16_t wcs[] = u"zß 水□"; //or "z\u00df\u6c34\U0001f34c"
    const size_t wcs_sz = sizeof wcs / sizeof *wcs;
    printf("%zu UTF-16 code units: [ ", wcs_sz);
    //输出：6 UTF-16 code units: [ 0x7a 0xdf 0x6c34 0xd83c 0xdf4c 0 ]
    for (size_t n = 0; n < wcs_sz; ++n)
        printf("%#x ", wcs[n]);
    printf("]\n");
    return 0;
}
```

25. 原始字符串文本（raw string literals）

原始字符串文本是指前缀包含 R 的字符串文本，该文本不会转义任何字符，这意味着分隔符"("和")"之间的任何内容都将成为字符串的一部分，\n、\t、\r 等都不会被处理。语法格式如下：

```
R "delimiter( raw_characters )delimiter"
```

这里 delimiter 是可选项。

【例 1-15】

```cpp
#include <iostream>
using namespace std;
int main(){
    string string1 = R"(Hello.\nWorld.\nC++.\n)";    //会输出：Hello.\nWorld.\nC++.\n
    string string2 = R"(\n\n\n\n)";                   //会输出：\n\n\n\n
    cout << string1 << endl;
    cout << string2 << endl;
    return 0;
}
```

1.2 C++14 新增的语言功能

C++14 新增了 8 个语言功能，其中的泛型 lambda 表达式（generic lambda expression）和 lambda 捕获初始化器（lambda capture initializer）放在第 2 章中介绍，其他 6 个新增的语言功能如下。

1. 二进制文本（binary literals）

在编写涉及数学计算或各种类型数字的程序时，我们通常喜欢用特定的前缀来指定每个数字类型，即十六进制数使用前缀 "0x"，八进制数使用前缀 "0"。现在，我们可以直接在 C++14 中编写二进制文本（形式为 0 和 1）。二进制文本可以以 0b 或 0B 为前缀。

【例 1-16】

```cpp
#include <iostream>
using namespace std;
int main(){
    int a = 0b00001111;       //二进制文本以"0b"开头
    cout << a << '\n';
    int b = 0B00001111;       //二进制文本以"0B"开头
    cout << b;
    return 0;
}
```

2. 变量模板（variable template）

变量模板可帮助开发人员轻松简单地定义参数化常量，以了解代码中使用的常量的类型和值。变量模板在处理数学常量、配置值或编译取决于类型或维度的常量时提供了灵活性。变量模板的语法格式如下：

```
template <parameter-list> variable-declaration
```

【例 1-17】

```cpp
#include <iostream>
using namespace std;
template <class T> constexpr T e = T(2.718281828459045);
class limits {
public:
    //静态数据成员模板声明
    template <typename T> static const T min;
```

```cpp
};
//静态数据成员初始化
template <typename T> const T limits::min = T(10.24);
int main(){
    cout << "Integer Type of e: " << e<int> << endl;
    cout << "Float Type of e: " << e<float> << endl;
    cout << limits::min<int> << endl;
    cout << limits::min<float>;
    return 0;
}
```

3. 返回类型推导（return type deduction）

C++14 使用 auto 返回类型，编译器将推断返回类型。在使用 lambda 表达式时，我们可以使用 auto 来推断返回类型，从而可以推断返回的引用或右值引用。例如：

```cpp
auto f(int i) {
    return i;
}
template <typename X>
auto& foo(X& x) {
    return x;
}
//返回一个约简类型的引用
auto a = [](auto& x) -> auto& { return foo(x); };
int b = 123;
int& c = a(b); //reference to b
```

4. 类型说明符 decltype(auto)

类型说明符 decltype(auto) 和 auto 一样可推断类型，但 decltype(auto) 会在保留其引用和 CV 限定符（const 和 volatile）的同时推断返回类型，而 auto 则不会。例如：

```cpp
const int x = 0;
auto x1 = x;                         //int
decltype(auto) x2 = x;               //const int
int y = 0;
int& y1 = y;
auto y2 = y1;                        //int
decltype(auto) y3 = y1;              //int&
int&& z = 0;
auto z1 = std::move(z);              //int
decltype(auto) z2 = std::move(z);    //int&&

auto f(const int& i){
    return i;                        //返回 int
}

decltype(auto) g(const int& i) {
    return i;                        //返回 const int&
}
```

5. 放宽对 constexpr 函数的限制（relaxing constraints on constexpr）

在 C++11 中，constexpr 函数只能包含一组非常有限的语法，包括（但不限于）typedef、using 和单个 return 语句。C++14 大大扩展了允许的语法集，包括最常见的语法，如 if 语句、多个返回、循环等。例如：

```
constexpr int factorial(int n){
    int ret = 1;
    for(int i=n; i>=1; i--)
    ret *= i;
    return ret;
}
```

6. [[deprecated]]属性

C++14 新增了[[deprecated]]属性，用于表示不建议使用某些函数、类等，并可能产生编译警告。如果提供了原因，则该原因将包含在编译警告中。例如：

```
[[deprecated]]void old_method();
[[deprecated("Use new_method instead")]]void legacy_method();
```

1.3 C++17 新增的语言功能

C++17 新增了 15 个语言功能，其中关于 lambda 表达式的语言新增功能为常量 lambda 表达式（constexpr lambda）和按值捕获 this 指针（lambda capture this by value），本书将在第 2 章介绍；关于模板的语言新增功能是折叠表达式（folding expression）、类模板参数推导（template argument deduction for class template），以及使用 auto 声明非类型模板参数（declaring non-type template parameters with auto），本书将在第 4 章中介绍。其他 10 个新增的语言功能如下所述。

1. 大括号初始化列表的自动推导新规则（new rules for auto deduction from braced-init-list）

在 C++17 之前的版本中，auto x{3}将被推导为一个 std::initializer_list<int>，现在将被推导为 int。在 C++17 中，当 auto 与变量一起进行初始化时，将自动推导数据类型。例如：

```
auto x1 {3};              //x1 的类型是 int
auto x2 {3.0};            //x2 的类型是 double
auto x3 = {1, 2, 3};      //x3 的类型是 std::initializer_list<int>
auto x4 {1, 2, 3};        //错误：不是单一元素
auto x5 = {1, 2.0, 3.0};  //错误：类型不一致
```

2. 内联变量（inline variable）

内联说明符 inline 不仅可用于变量和函数，内联声明的变量与内联声明的函数具有相同的语义；还可用于声明和定义静态成员变量，这样就不需要在源文件中对静态成员变量进行初始化。

【例 1-18】

```
#include <iostream>
using namespace std;
inline int inline_var = 20;

class myClass {
```

```
public:
    inline static int var = 10;          //使用 inline 直接初始化静态成员变量
};
int main(){
cout << inline_var;
    cout << myClass::var;                //访问静态成员变量
    return 0;
}
```

3．嵌套命名空间（nested namespace）

使用命名空间解析运算符可创建嵌套命名空间。

【例 1-19】

```
#include <iostream>
using namespace std;
namespace A{
    void f(){
        cout << "Inside namespace A" << endl;
    }
    namespace B{
        void f(){
            cout << "Inside namespace B" << endl;
        }
    }
}
using namespace A::B;
int main (){
    f();                                 //调用 namespace B 的 f()
    return 0;
}
```

4．结构化绑定（structured binding）

结构化绑定用于将指定的名称绑定到初始值设定项的子对象或元素，其语法格式为：

```
auto [x, y, z] = expr;
```

其中 expr 的类型是一个类似元组的对象，其元素将绑定到变量 x、y 和 z。类似元组的对象包括 std::tuple、std::pair、std::array 和 aggregate 结构。

绑定数组的示例如下：

```
int a[2] = {1, 2};
auto [x, y] = a;          //创建 e[2]，然后将 a 复制到 e，x 引用 e[0]，y 引用 e[1]
auto& [xr, yr] = a;       //xr 引用 a[0]，yr 引用 a[1]

float x{};
char y{};
int z{};
std::tuple<float&, char&&, int> tpl(x, std::move(y), z);
const auto& [a, b, c] = tpl;
```

5. 带初始化的选择语句（selection statements with initializer）

C++17 中的 if 和 switch 语句可简化为常用代码模式并帮助用户保持严格的范围，其语法格式如下：

```cpp
if (init; condition)
switch (init; condition)
if (unsigned i = std::rand(); i % 2 == 0) {
    std::cout << "i is even" << std::endl;
}
else {
    std::cout << "i is odd" << std::endl;
}
if(const auto [x,y,z] = func_returning_a_tuple(); x + y < z){
    ... //执行相关语句
}
```

6. 常量 if 语句

C++17 新增的 if constexpr 功能允许基于常量表达式的分支编译。与在运行时计算常量的 if 语句不同，if constexpr 功能允许编译器丢弃不适用的分支代码，这意味着在编译过程中，仅编译条件为 true 的分支代码，而丢弃其他分支代码。if constexpr 的语法格式如下：

```cpp
if constexpr (condition) {
    ... //判断条件为 true 时编译这段代码
} else {
    ... //判断条件为 false 时编译这段代码
}
```

【例 1-20】

```cpp
#include <iostream>
using namespace std;
template <typename C>
void printInfo(const C& value) {
    if constexpr (is_integral_v<C>) {
        cout << "Integer Value " << value << endl;
    }
    else {
        cout << "Non-Integer value:" << value << endl;
    }
}
int main() {
    printInfo(10);
    printInfo(3.15);
    return 0;
}
```

7. UTF-8 字符文本（UTF-8 character literals）

以 u8 开头的字符文本是 char 类型的字符文本。UTF-8 字符文本的值等于其在 ISO 10646 中的码位值。对于 UTF-8 字符串文本，其文本的类型为 char[N]，N 是 UTF-8 编码单元的字符串大小，数组中的每个字符元素都使用 UTF-8 编码。例如：

```
char x = u8'x';
char8_t s2[] = u8"a 猫□";
```

【例 1-21】

```cpp
#include <iostream>
int main(){
    std::cout << "Unicode: " << '\u03C0' << '\n';          //希腊字母 pi
    std::cout << "Beyond BMP: " << '\U0001F600' << '\n';   //狞笑表情
    std::cout << "UTF-8: " << u8"\u03BB" << '\n';          //希腊字母 lambda
    return 0;
}
```

8. 枚举的直接列表初始化（direct-List-Initialization of enum）

枚举类型在 C++17 中可以使用花括号{}进行初始化，例如：

```cpp
enum byte : unsigned char {};
byte a {0};              //正确
byte b {-1};             //错误
byte c = byte{1};        //正确
byte d = {42};           //错误
byte e = byte{42};       //正确
```

9. [[fallthrough]]、[[nodiscard]]和[[maybe_unused]]属性

C++17 新增了三个新属性：[[fallthrough]],[[nodiscard]],[[maybe_unused]]。

（1）[[fallthrough]]属性：用于向编译器指示在 switch 语句中失败是预期行为。例如：

```cpp
switch (n) {
    case 1: [[fallthrough]]
    //...
    case 2:
    //...
    break;
}
```

（2）[[nodiscard]]属性：当函数或类具有此属性并且其返回值被丢弃时发出警告。例如：

```cpp
[[nodiscard]] bool do_something() {
    return is_success;   //true 表示成功，false 表示失败
}
do_something();
```

（3）[[maybe_unused]]属性：用于向编译器指示变量或参数可能未使用并且是预期的。例如：

```cpp
void my_callback(std::string msg, [[maybe_unused]] bool error){
    log(msg);
}
```

10. 运算符__has_include

运算符__has_include 可用于#if 和#elif 表达式，用来检查头文件或源文件（操作数）是否可被包含。其中的一个情况是使用两个工作方式相同的库，如果在系统上找不到首选的库，则使用备份/实验库。例如：

```
#ifdef __has_include
#  if __has_include(<optional>)
#     include <optional>
#     define have_optional 1
#  elif __has_include(<experimental/optional>)
#     include <experimental/optional>
#     define have_optional 1
#     define experimental_optional
#  else
#     define have_optional 0
#  endif
#endif
```

1.4 C++20 新增的语言功能

C++20 新增了十几项语言功能，本书将在第 2 章中介绍 lambda 模板化语法（template syntax for lambdas）、lambda 参数包捕获（lambda capture of parameter pack）和不推荐隐式捕获 this 指针（deprecate implicit capture of this），在第 3 章中介绍 constexpr 虚函数（constexpr virtual function），在第 4 章中介绍非类型模板参数中的类类型（class type in non-type template parameter），其他的语言新增功能在本节介绍。

1. 协程（coroutine）

协程是一个特殊函数，它可以暂停执行和恢复执行。若要定义协程函数，则必须将 co_return、co_await 或 co_yield 关键字放在函数主体中。C++的协程是无栈的，除非编译器进行优化，否则将协程的状态分配在堆上。

下面是一个协程示例，它是生成器函数，在每次调用时生成一个值。

```
generator<int> range(int start, int end){
    while (start < end) {
        co_yield start;
        start++;
    }
    //co_return;                  //隐式返回 co_return
}
for (int n : range(0, 10)) {
    std::cout << n << std::endl;
}
```

上面的生成器函数生成从开始到结束（不含结束）的值，每个迭代步骤都会产生存储在 start 中的当前值。生成器函数在每次调用时保持其状态不变（在本例中，调用是针对 for 循环中的每次迭代的）。co_yield 用于获取给定的表达式，生成（即返回）表达式的值，并在该点挂起协程。恢复后，co_yield 后继续执行。

协程的另一个示例是任务，可在等待时执行异步计算，例如：

```
task<> tcp_echo_server(){
    char data[1024];
    while (true)
```

```
        {
            std::size_t n = co_await socket.async_read_some(buffer(data));
            co_await async_write(socket, buffer(data, n));
        }
}
```

上面的代码引入了关键字 co_await，该关键字采用了一个表达式。如果正在等待的内容（在上面的代码中为读取或写入）尚未准备就绪，则暂停执行，否则继续执行。请注意，在有钩子函数的情况下，co_yield 使用 co_await。

2．概念（concept）

概念被命名为编译时谓词（predicate），用于约束类型，可以采用以下形式：

```
template < template-parameter-list >
concept concept-name = constraint-expression;
```

其中 constraint-expression 的计算结果为 constexpr 布尔值。约束应对语义要求进行建模，如类型是数字类型还是可哈希类型。如果给定的类型不满足其绑定的概念（即 constraint-expression 返回 false），则会导致编译器错误。由于约束是在编译时评估的，因此可以提供更有意义的错误消息和运行时安全性。例如：

```
//约束表达式对 T 没有限制
template <typename T>
concept always_satisfied = true;

//限制 T 为整型
template <typename T>
concept integral = std::is_integral_v <T>;

//限制 T 为整型和有符号类型
template <typename T>
concept signed_integral = integral<T> && std::is_signed_v<T>;

//限制 T 为整型和无符号类型
template <typename T>
concept unsigned_integral = integral<T> && !signed_integral<T>;
```

3．指定初始化（designated initializer）

采用指定初始化功能，可以指定初始值设定项列表中未显式列出的任何成员字段都是默认初始化的。例如：

```
struct A {
    int x;
    int y;
    int z = 123;
};
A a {.x = 1, .z = 2};    //a.x = 1, a.y = 0, a.z = 2
struct Date{
    int year;
    int month;
    int day;
```

```
};
Date date{.year = 2024, .month = 5, .day = 20 };
```

4．带初始化的基于范围的 for 循环（range-based for loop with initializer）

该功能可简化常见的代码模式，有助于保持作用域的紧凑，并为常见的生存期问题提供了优雅的解决方案。例如：

```
std::array data = {"hello", ",", "world"};
for (std::size_t i = 0; auto& d : data) {
    std::cout << i++ << ' ' << d << '\n';
}
for (auto v = std::vector{1, 2, 3}; auto& e : v) {
    std::cout << e;
}
```

5．[[likely]]和[[unlikely]]属性

[[likely]]和[[unlikely]]属性可提示编译器被标记的语句，具有很高的执行概率。例如：

```
switch (n) {
case 1:
    //...
    break;
[[likely]] case 2:          //n == 2 比其他值有较高可能性
    //...
    break;
}
```

如果 if 语句的右括号后出现一个[[likely]]或[[unlikely]]，则表示分支可能或不太可能执行其子语句。例如：

```
int random = get_random_number_between_x_and_y(0, 3);
if (random > 0) [[likely]] {
    //...
}
```

[[likely]]和[[unlikely]]属性也可以应用于迭代语句，例如：

```
while (unlikely_truthy_condition) [[unlikely]] {
    //...
}
```

6．说明符 explicit（bool）

在编译时可以有条件地选择构造函数是否为显式构造函数，如果是显式构造函数，则判断该显式构造函数是否与指定的 explicit(true)相同。

```
struct foo {
    //指定非整型（string、float 等）要求显式构造函数
    template <typename T>
    explicit(!std::is_integral_v<T>) foo(T) {}
};
foo a = 123; //OK
foo b = "123";            //错误：显式构造函数不是候选项（explicit 说明符的计算结果为 true）
```

```cpp
foo c {"123"};                //OK
template <typename T>
struct Foo {
    explicit Foo(T) {}
};
template <typename T>
struct Pair {
    template <typename U>
    Pair(U, U) {}
};
Pair<Foo<int>> pair = {1, 2}; //正确
Foo<int> foo1 = 1; //错误：从 int 到 Foo<int>的转换不可行
Foo<int> foo2 = 2; //错误：从 int 到 Foo<int>的转换不可行
```

7．即时函数（immediate function）

在 C++20 中，即时函数是指每次调用该函数时都会直接或间接生成编译时常量表达式的函数。即时函数通过在其返回类型之前使用 consteval 关键字进行声明。与 constexpr 函数类似，但即时函数具有 consteval 说明符，且必须生成常量。例如：

```cpp
consteval int sqr(int n) {
    return n * n;
}
constexpr int r = sqr(100);          //正确
int x = 100;
int r2 = sqr(x);    //错误：x 的值不可以用在常量表达式中，但使用 constexpr 函数就是正确的
```

8．引入枚举作用域（using enum）

引入枚举作用域可以将枚举的成员纳入一定的范围，以提高可读性。对比以下代码：

（1）未采用引入枚举作用域功能的格式：

```cpp
enum class rgba_color_channel { red, green, blue, alpha };
std::string_view to_string(rgba_color_channel channel) {
    switch (channel) {
        case rgba_color_channel::red:   return "red";
        case rgba_color_channel::green: return "green";
        case rgba_color_channel::blue:  return "blue";
        case rgba_color_channel::alpha: return "alpha";
    }
}
```

（2）采用引入枚举作用域功能的格式：

```cpp
enum class rgba_color_channel { red, green, blue, alpha };
std::string_view to_string(rgba_color_channel my_channel) {
    switch (my_channel) {
        using enum rgba_color_channel;
        case red:   return "red";
        case green: return "green";
        case blue:  return "blue";
        case alpha: return "alpha";
```

```
        }
}
```

9. char8_t

char8_t 用于声明 UTF-8 无符号字符型的变量或函数，与无符号整型相同。

【例 1-22】

```
#include <uchar.h>
#include <stdio.h>
int main(void){
    char8_t str[] = u8"zß 水 □"; //or "z\u00df\u6c34\U0001f34c"
    size_t str_sz = sizeof(str);
    printf("%zu UTF-8 code units: [ ", str_sz);
    for (size_t n = 0; n < str_sz; ++n)
    printf("%02X ", str[n]);
    printf("]\n");
}
```

10. 关键字 constinit

关键字 constinit 用于确保变量在编译时或程序启动时完成静态初始化，即零初始化和常量初始化。

【例 1-23】

```
#include <iostream>
using namespace std;
constinit int x = 42;              //声明一个 constinit 变量
int main(){
    cout << "x = " << x << endl;
    return 0;
}
```

11. 关键字 consteval

C++20 引进了另一个关键字 consteval，该关键字只能用于函数，不能用于变量，并强制所有的函数调用都是在编译时发生的。

【例 1-24】

```
consteval int sum(int a, int b){
    return a + b;
}
constexpr int sum_c(int a, int b){
    return a + b;
}
int main() {
    constexpr auto c = sum(100, 100);
    static_assert(c == 200);
    constexpr auto val = 10;
    static_assert(sum(val, val) == 2*val);
    int a = 10;
    int b = sum_c(a, 10);              //正确
    //int d = sum(a, 10);               //错误：a 的值不能用于常量表达式
```

```
    return 0;
}
```

关键字 const、constexpr、consteval 和 constinit 的应用场景如表 1.5-1 所示。

表 1.5-1 关键字 const、constexpr、consteval 和 constinit 的应用场景

关键字	是否用于 auto 变量	是否用于静态/线程变量	是否用于函数	是否用于常量表达式
const	Y	Y	const 成员函数	有时
constexpr	Y	Y	constexpr 函数	Y
consteval	N	N	consteval 函数	Y
constinit	N	强制常量初始化	N	N

(1) 关键字 consteval 和 constexpr 的区别。
- consteval 只能用于函数和常量表达式,不能用于变量。
- constexpr 既可用于函数,也可用于变量。
- consteval 在编译时进行函数评估。
- constexpr 既可在编译时执行函数,也可以在运行时执行函数(作为常规函数)。

(2) 关键字 const 和 constexpr 的区别。

const 是 C++早期就开始使用的关键字,不仅可用于对象,以指示其不变性;还可用于非静态成员函数,从而在给定类型的常量实例上调用这些函数。const 并不意味着任何"编译时"评估,尽管编译器可以优化代码并这样做,但通常是在运行时初始化的。例如:

```
const int count = 3;                        //正确
std::array<double, count> doubles {1.1, 2.2, 3.3};
```

const 不能用于 double 类型的变量,例如:

```
const double dCount = 3.3;                  //错误
std::array<double,static_cast<int>(dCount)>moreDoubles{1.1, 2.2, 3.3};
```

constexpr 是 C++11 引入的一个新关键字,它进一步推动了对可用于常量表达式的变量和函数的控制。现在,constexpr 不是 C++的一个技巧或特例,而是一个完整、更易于理解的解决方案。现在可以这么使用:

```
constexpr double dCount = 3.3;
std::array<double, static_cast<int>(dCount)> doubles2 {1.1, 2.2, 3.3};
```

【例 1-25】

```
#include <array>
#include <cstdlib>
struct Point{
    int x {0};
    int y {0};
    constexpr int dist(const Point& other) const{
        return sqrt((x - other.x)*(x - other.x) + (y - other.y)*(y - other.y));
    }
};
int main(){
```

```
    constexpr Point a {0, 0};
    constexpr Point b {10, 12};
    static_assert(a.dist(b) >= 10.0);
    Point c {100, 1};
    Point d {10, 11};
    return c.dist(d);
}
```

总结：
- const 可用于各种对象，以指示它们的不变性。
- 整型常量在初始化后可用于常量表达式。
- 根据定义，constexpr 可用于常量表达式。
- constexpr 可用于函数，以表明可以调用函数来生成常量表达式。
- const 可用于成员函数，以指示函数不更改数据成员。

（3）关键字 constexpr 和 constinit 的区别。

关键字 constinit 用于强制对静态变量或线程局部变量执行静态初始化。与 const 或 constexpr 不同，constinit 并不意味着对象是不可变的。更重要的是，constinit 不能用于常量表达式，这就是为什么在下面的例子中，不能用 global 启动 another 或者使用 global 作为数组大小的原因。

【例 1-26】

```
#include <array>
constexpr int compute(int v) { return v*v*v; }
constinit int global = compute(10);
//constinit int another = global;          //不正确
int main() {
    global = 100;
    //global 不是一个整型常量，不能作为数组的大小
    //std::array<int, global> arr;
    return 0;
}
```

总结：
- constexpr 变量是常量，可用于常量表达式。
- constinit 变量不是常量，不能在常量表达式中使用。
- constexpr 可用于局部自动变量，这在 constinit 中是不允许的；constinit 只能用于静态或 thread_local 对象。
- 可以使用 constexpr 声明函数，但不能使用 constinit 声明函数。

1.5 本章小结

C++11 是现代 C++ 的主要版本。从 C++11 起，C++ 新增了大量功能。现代 C++ 的新增功能分为语言新增功能和 STL 新增功能。本章主要介绍 C++11/14/17/20 等的语言新增功能。对于给出的每一项语言新增功能，本章都给出了功能介绍和示例代码。

C++14 是 C++11 之后的一个重要版本，主要对 C++11 做了一些改进和缺陷修复。在语言新

增功能方面，C++14 新增了 8 项语言功能。

C++17 是一次较大的修改。一方面，一些新增的语言功能从 C++17 中删除了（如 std::auto_ptr 等），也废弃了一些语言功能（如 std::result_of 等）；另一方面，又新增了很多语言功能，主要集中在常量表达式、变量、模板、命名空间和新属性等方面，本章对其中的绝大多数语言新增功能做了详细介绍，并给出了示例代码。

C++20 是 C++的全新版本，现在主流的编译器都支持 C++20。C++20 的变化是巨大的，它将比 C++11 更显著地改变编程方式。C++20 新增了十几项语言功能，增加了更多的 STL 功能。本章对 C++20 的部分语言新增功能进行了详细介绍。

在介绍 C++不同版本的语言新增功能后，本章对比了关键字 const、constexpr、consteval 和 constinit 的应用场景，目的是帮助读者加强对这些关键字的理解。

第 2 章 lambda 表达式

C++11 新增了 lambda 表达式。lambda 表达式可以被看成匿名函数，这些函数用于不会被重用的代码片段，因此不需要名称。

2.1 C++11 中的 lambda 表达式

lambda 表达式语法格式如下：

[captures](params) mutable exception trailing-type {body};

其中：

（1）captures 是捕获子句，可通过引用捕获、值捕获和混合捕获等方式从封闭的范围内捕获外部变量。用于捕获变量的语法如下：
- [&]：通过引用捕获方式捕获所有的外部变量。
- [=]：通过值捕获方式捕获所有的外部变量。
- [ ]：不访问本地变量。
- [this]：在成员函数中，可以直接捕获 this 指针。

例如：[=, &a]可通过引用捕获方式来捕获变量 a，通过值捕获方式捕获其他的局部变量；[&, a]可通过值捕获方式捕获变量 a，通过引用捕获方式捕获所有的局部变量；[a, &b]可通过值捕获方式捕获变量 a，通过引用捕获方式捕获变量 b。

（2）params：参数列表，可选项，参数之间用逗号隔开，参数将在 body 部分使用。

（3）mutable：可选项，允许在 body 部分修改按值捕获的变量，并调用变量的非常量成员函数。

（4）exception：可选项，提供动态的异常说明。

（5）trailing-type：可选项，尾随类型。如果未提供尾随类型，则 lambda 表达式返回类型的确定规则如下：
- 如果 body 仅由返回表达式 return 语句组成，则返回类型是该表达式在左值到右值转换、数组到指针转换和函数到指针转换后的类型；否则，返回类型为 void。
- 如果尾随类型是-> auto，则自动推导返回类型。

（6）body：函数体。lambda 表达式的函数体是复合语句，可以包含任何普通函数或成员函数中的任何内容。普通函数和 lambda 表达式的函数体都可以访问以下类型的变量：
- 在封闭作用域中捕获的变量。
- 参数。

- 本地声明的变量。
- 如果类数据成员是在类内部声明的,则将被捕获。
- 任何具有静态存储持续时间的变量,如全局变量。

lambda 表达式在内部是如何工作的呢?我们先看一个简单的 lambda 表达式。

```
[&i]( ){ std::cout << i;}
```

这个 lambda 表达式等价于:

```
struct anonymous{
    int &m_i;
    anonymous(int &i) : m_i(i) {}
    inline auto operator()() const {
        std::cout << i;
    }
};
```

编译器为每个 lambda 表达式生成唯一的匿名函数(也称为匿名闭包)。捕获列表将成为匿名函数中的构造函数参数。如果将参数捕获为值,则会在匿名函数中创建相应类型的数据成员。此外,程序员可以在 lambda 表达式的匿名函数参数中声明变量,该参数将成为函数调用运算符 operator()的参数。明白了 lambda 表达式的工作机理,对我们理解使用 lambda 表达式非常有用。

我们来看一个简单 lambda 表达式的示例代码。

【例 2-1】

```
#include <vector>
#include <stream>
using namespace std;

int main(){
    vector<int> v1 = {5, 3, 8, 2};
    vector<int> v2 = {8, 4, 9, 1, 3};
    //通过引用捕获方式捕获变量 v1 和 v2
    auto append = [&] (int a){
        v1.push_back(a);
        v2.push_back(a);
    };
    //现在将 12 和 15 添加到变量 v1 和 v2 的尾部
    append(12);
    append(15);
    //通过值捕获方式来捕获变量 v1
    [v1](){
        for (auto t = v1.begin(); t != v1.end(); t++){
            cout << *t << " ";
        }
    };
    return 0;
}
```

在上面的例子中,我们应用了引用捕获方式[&]和值捕获方式[v1]。当需要改变被引用变量时,采用引用捕获方式[&]，否则采用值捕获方式[v1]。上面的第 1 个 lambda 表达式有一个参数，所以在调用时需要传入实参，第 2 个 lambda 表达式没有参数。

下面的例子使用空捕获子句[ ]，auto 类型参数，并带有尾随类型，它常用于算法函数中的比较函数。

【例 2-2】

```cpp
#include <algorithm>
#include <iostream>
#include <string>
#include <vector>

void printArray(auto& A){
    for (auto a : A)
    cout << a << " " << endl;
}

int main(){
    //定义一个 lambda 表达式并存放在 greater 中，注意该 lambda 表达式是带返回类型的
    auto greater = [](auto a, auto b) -> bool {
        return a > b;
    };

    vector<int> v0{ 1, 5, 3, 8, 6, 72, 146 };
    vector<double> v1{ 22.45, 41.33, 32.68, 39.24, 25.58 };
    vector<string> v2{ "Jack", "Hansen", "Roy", "David" };

    sort(v0.begin(), v0.end(), greater );
    printArray(v0);

    sort(v1.begin(), v1.end(), greater );
    printArray(v1);

    sort(v2.begin(), v2.end(), greater );
    printArray(v2);
    return 0;
}
```

由于 lambda 表达式是类型化的，因此可以将其赋值给 auto 变量或者函数对象。上面的例子都是将 lambda 表达式赋值给 auto 变量。下面的例子将 lambda 表达式赋值给函数对象。

【例 2-3】

```cpp
#include <functional>
#include <iostream>

int main() {
    int i = 3;
    int j = 5;

    std::function<int(void)> f = [i, &j]{return i+j;};
```

```
        i = 22;
        j = 44;
        cout << f() << endl;
        //输出结果为47,因为j是通过引用捕获方式捕获的,i是通过值捕获方式捕获的
        return 0;
}
```

下面的例子在 lambda 表达式中使用 mutable 和抛出异常。

【例 2-4】

```
#include <iostream>
using namespace std;

int main() {
    int x = 0;
    int y = 0;
    [&, y] (int a) mutable { x = ++y + a; }(9);
    cout << x << " " << y << endl;              //输出为 10 和 0
}
```

在捕获部分中,y 是通过值捕获方式捕获的,因此在调用 lambda 表达式后,y 的值依然是 0。添加 mutable 后,可以在函数体里进行 ++y 操作。需要记住:x 是通过引用捕获方式捕获的,在调用 lambda 表达式后 x 的值是 10,y 的值还是 0。尾部的(9)表示调用该 lambda 表达式时传入给 a 的实参是 9。

在下面的例子中,我们看到的 lambda 表达式是"[=]() mutable throw() -> int{}",即采用值捕获方式、无参数、可变异、抛出异常、带返回类型。

【例 2-5】

```
#include <algorithm>
#include <iostream>
#include <vector>
#include <string>

template <typename T>
void output(const string& str, const T& t){
    std::cout << str;
    for (const auto& a : t) {
        std::cout << a << " ";
    }
    std::cout << endl;
}

void initVector(vector<int>& v){
    //定义一个静态局部整型变量
    static int val = 0;
    std::generate(v.begin(), v.end(), []{return ++val;});
}
```

```cpp
int main(){
    const int count = 9;
    vector<int> v(count, 1);

    int a = 1;
    int b = 1;

    //lambda 表达式的函数体中，当前元素是前两个元素之和
    std::generate_n(v.begin() + 2,
    count - 2,
    [=]() mutable throw() -> int {
        int c = a + b;
        a = b;
        b = c;
        return c;
    });

    //v 的值为: 1 1 2 3 5 8 13 21 34
    output("vector v after call to generate_n() with lambda:", v);
    std::cout<<"a: "<< a <<" b: "<< b <<endl;        //记住：a 和 b 的值还是 1

    initVector(v);           //v 的值为: 1 2 3 4 5 6 7 8 9
    output("vector v after 1st call to fillVector(): ", v);
    initVector(v);           //v 的值为: 10 11 12 13 14 15 16 17 18
    output("vector v after 2nd call to fillVector(): ", v);
    return 0;
}
```

下面给出的是立即调用 lambda 表达式的示例。

【例 2-6】

```cpp
#include <iostream>
using namespace std;

int main() {
    int n = [](int x, int y){ return x + y; }(5, 4);
    cout << n << endl;
    return 0;
}
```

当 lambda 表达式用于类成员函数中，可以通过捕获 this 指针来使用类成员变量，也可以使用 [=]和[&]来捕获成员变量。

```cpp
class myClass{
public:
    void foo(int i);
    void bar(int);
};

void myClass::foo(int i){
```

```
        [&] {};                 //正确：通过引用捕获方式捕获
        [&, i] {};              //正确：通过引用捕获方式捕获，i 是通过值捕获方式捕获的
        [&, &i] {};             //错误：主捕获方式是引用捕获方式，其他不能再使用引用捕获方式
        [&, this] {};           //正确：等价于[&]
        [&, this, i] {};        //正确：等价于[&, i]
}

void myClass::bar(int i){
        [=] {};                 //正确：通过值捕获方式捕获
        [=, &i] {};             //正确：通过值捕获方式捕获，i 是通过引用捕获方式捕获的
        [=, *this]{};           //在 C++17 以前是错误的，无效的语法，在 C++17 中是正确的
        [=, this] {};           //在 C++20 以前是错误的，在 C++20 中是正确的
}
```

当捕获的变量带有初始值时，就像声明并显式捕获带有初始值的 auto 类型变量一样，该变量的声明区域是 lambda 表达式的主体。例如，下面的 lambda 表达式中的 r。

```
int x = 4;
auto y = [&r = x, x = x + 1]() -> int{
        r += 2;                 //r 是 x 的引用，所以 x 为 6
        return x * x;           //这里的 x 是通过值捕获方式捕获的，因此是 5
}();                            //直接调用，结果 x = 6，y = 25
```

2.2 C++14 对 lambda 表达式的扩展

C++14 对 lambda 表达式的扩展有以下三点：
- lambda 捕获初始化（lambda capture initializer）。
- 泛型 lambda 表达式（generic lambda expression）。
- 默认参数（default parameter）。

下面我们分别介绍每一种扩展。

2.2.1 lambda 捕获初始化

lambda 捕获初始化是指在捕获子句中引入新的变量并进行初始化，而这些变量无须在 lambda 表达式的封闭作用域中。初始化可以表示为任意表达式，新变量的类型是从 lambda 表达式生成的类型中推导出来的。此功能允许程序从周围范围捕获仅移动变量（如 std::unique_ptr）并在 lambda 表达式中使用它们。

```
int factory(int i){
        return i * 10;
}

auto f = [x = factory(2)] { return x; };        //对捕获变量 x 进行初始化

auto generator = [x = 0] () mutable {
        //这里必须加上 mutable，因为 x 现在已经改变了
```

```
        return ++x;
    };
    auto a = generator();
```

现在可以使用 std::move（或 std::forward）将以前只能通过复制或引用捕获方式捕获的变量引入 lambda 表达式中。例如：

```
auto p = std::make_unique<int>(1);              //p 为 std::unique_ptr 类型的智能指针
auto p1 = [=]{ *p = 5; };                       //错误：std::unique_ptr 不能被拷贝

//请注意：捕获子句中的 p 是 lambda 表达式中私有的新变量，不是引用原始的 p
auto p2 = [p = std::move(p)] { *p = 5; };       //原来的 p 被清空
```

为了避免上面的混淆，在使用引用捕获方式时，可以使用不同的名称，而非被引用捕获的变量名。例如：

```
auto x = 1;
auto f =[&r = x, x = x * 10]()->int{
    ++r;                    //注意 r 是 x 的引用，因此 r 是 2
    return r + x;           //这里的 x 是通过值捕获方式捕获的，因此是 10
};
int y = f();                //x 为 2，y 为 12
```

2.2.2 泛型 lambda 表达式

C++14 引入了泛型 lambda 表达式，使 lambda 表达式呈现多态性。具体表现为：可以在 lambda 表达式的输入参数中使用关键字 auto，这样编译器就可以在编译时推断输入参数的类型。对于类型为 auto 的输入参数，将按照出现顺序制定参数模板，并将参数添加到参数模板中。例如：

```
auto identity = [](auto x) { return x; };
int three = identity(3);
std::string s = identity("foo");

auto glambda = [](auto a, auto&& b) { return a < b;};
bool b = glambda(3, 3.14);                       //OK
```

在泛型 lambda 表达式中，auto 类型的输入参数可以是参数包，这是因为需要调用的函数参数是参数包。例如下面的嵌套式 lambda 表达式：

```
auto vglambda = [](auto printer){                //泛型 lambda 表达式
    return [=](auto&&... ts)                     //泛型 lambda 表达式，ts 是参数包。
    {
        printer(std::forward<decltype(ts)>(ts)...);
        return [=] { printer(ts...); };
    };
};

auto p = vglambda([](auto v1, auto v2, auto v3){
    std::cout << v1 << v2 << v3;});
p(1, 'a', 3.14);                                 //输出为 1a3.14
```

2.2.3 默认参数

从 C++14 起，lambda 表达式可以带默认值参数，例如：

```
auto func = [](int i = 5) { return i + 4; };
std::cout << "func: " << func() << '\n';          //调用时参数使用默认值
```

下面的例子把 lambda 表达式放入函数模板中，采用的捕获方式为引用捕获。

【例 2-7】

```
#include <vector>
#include <algorithm>
#include <iostream>
using namespace std;

//给 vector 中的每一个元素加上负号
template <typename T>
void sign_all(vector<T>& v) {
    for_each(v.begin(), v.end(), [](T& n) { n = -n;});
}


template <typename T>
void output_all(const vector<T>& v){
    for_each(v.begin(), v.end(), [](const T& n) { cout << n << " ";});
}

int main() {
    vector<int> v;
    v.push_back(18);
    v.push_back(-36);
    v.push_back(28);
    v.push_back(-41);
    output_all(v);              //输出为 18   -36   28   -41
    sign_all(v);
    cout << "After sign_all():" << endl;
    output_all(v);              //输出为-18   36   -28   41
    return 0;
}
```

下面的例子把 lambda 表达式放入类的成员函数，并在 lambda 表达式中访问类的成员变量。

【例 2-8】

```
#include <algorithm>
#include <iostream>
#include <vector>
using namespace std;

class Amplifer {
public:
```

```cpp
        explicit Amplifer(float a) : _amplifer(a) {}
        void DoAmplify(const vector<float>& v) const {
            for_each(v.begin(), v.end(), [=](float n) { cout << n *
            _amplifer << endl; });
        }
private:
    float _amplifer;
};

int main(){
    vector<float> v;
    v.push_back(10.0);
    v.push_back(20.0);
    v.push_back(30.0);
    v.push_back(40.0);

    Amplifer a(2.5);
    a.DoAmplify(v);
    return 0;
}
```

接下来我们看一个高阶返回的 lambda 表达式示例。

【例 2-9】

```cpp
#include <iostream>

const auto less_than = [](auto x) {
    return [x](auto y) {
        return y < x;
    };
};

int main(void){
    auto less_than_five = less_than(5);
    std::cout << less_than_five(3) << std::endl;
    std::cout << less_than_five(10) << std::endl;
    return 0;
}
```

下面的例子将 lambda 表达式赋值给 std::function 对象，通过递归调用求斐波那契序列。

【例 2-10】

```cpp
#include <algorithm>
#include <functional>
#include <iostream>
#include <vector>
int main(){
    auto nth_fibonacci = [](int n){
        std::function<int(int, int, int)> fib = [&](int n, int a, int b)
        {
```

```
                return n ? fib(n - 1, a + b, a) : b;
            };
            return fib(n, 0, 1);
        };
        for (int i{1}; i <= 8; ++i)
            std::cout << nth_fibonacci(i) << (i < 8 ? ", " : "\n");
        return 0;
    }
```

2.3 C++17 对 lambda 表达式的扩展

C++17 对 lambda 表达式的扩展有两项：
- 常量 lambda 表达式（constexpr lambda）。
- 按值捕获 this 指针（lambda capture this by value）。

2.3.1 常量 lambda 表达式

常量 lambda 表达式是指在编译 lambda 表达式时使用常量表达式。当常量表达式允许捕获 lambda 表达式或者对引入的每个数据成员进行初始化时，可以将 lambda 表达式声明为常量表达式。例如：

```
int y = 5;
auto answer = [y]() constexpr {
    int x = 10;
    return y + x;
};
static_assert(answer() == 15);

constexpr int increment(int n){
    return [n] { return n + 1; }();
}
static_assert(increment(1) == 2);

auto identity = [](int n) constexpr {return n;};
static_assert(identity(123) == 123);

constexpr auto add = [](int x, int y) {
    auto L = [=] { return x; };
    auto R = [=] { return y; };
    return [=] { return L() + R(); };
};

static_assert(add(1, 2)() == 3);

constexpr int addOne(int n) {
    return [n]{return n + 1;}();
```

```
}
static_assert(addOne(1) == 2);
```

2.3.2 按值捕获 this 指针

lambda 表达式可捕获 this 指针，这使我们能够在 lambda 表达式中访问类的成员变量和成员函数。在 C++17 以前的版本中，我们只能通过引用捕获方式来捕获 this 指针，现在可以按值捕获 this 指针。当引用的对象是临时的或超出范围时，这会导致引用悬而未决的问题。按值捕获 this 指针可避免该问题。

【例 2-11】

```cpp
#include <iostream>
using namespace std;

class Counter {
public:
    Counter(): count(0){}
    void Increment(){
        //定义一个 lambda 表达式，按值捕获 this 指针
        auto incrementLambda = [*this]() mutable { //使用 mutable，因为需要修改成员变量 count
            count++;
        };

        //调用 lambda 表达式
        incrementLambda();
    }
    int GetCount() const { return count; }

private:
    int count;
};

int main(){
    Counter counter;
    counter.Increment();
    counter.Increment();
    cout << "Count: " << counter.GetCount() << endl;
    return 0;
}
```

【例 2-12】

```cpp
#include <iostream>
using namespace std;

struct MyObj{
    int value {123};
    auto getValueCopy() {
        return [*this] { return value; };          //C++17 允许按值捕获 this 指针
```

```cpp
    }
    auto getValueRef() {
        return [this] { return value; };          //在 C++11 中通过引用捕获方式捕获 this 指针
    }
};

int main(){
    MyObj mo;
    auto valueCopy = mo.getValueCopy();
    auto valueRef = mo.getValueRef();
    mo.value = 321;
    int a = valueCopy();                          //123
    int b = valueRef();                           //321
    cout << a << " " << b << endl;
    return 0;
}
```

2.4 C++20 对 lambda 表达式的扩展

C++20 对 lambda 表达式的扩展较 C++14 和 C++17 相对较少，主要有以下两点：
- lambda 模板语法（template syntax for lambda）。
- lambda 参数包捕获（lambda capture of parameter pack）。

2.4.1 lambda 模板语法

lambda 模板语法可对输入参数的类型进行模板化，例如：

```cpp
auto glambda = []<class T>(T a, auto&& b){ return a < b; };
auto ylambda = []<class T>(const T& a, const T& b) { return a+b;}
auto myLambda = []<class T>(T&& v){ std::cout << v << '\n';};
```

2.4.2 lambda 参数包捕获

lambda 参数包捕获包括值捕获和引用捕获两种方式。通过值捕获 lambda 参数包的示例如下：

```cpp
template<typename... Ts>
void f(Ts&&... args)                    //函数的参数包
{
    auto myLambda = [...args = std::forward<Ts>(args)](){};
}
```

通过引用捕获 lambda 参数包的示例如下：

```cpp
template<typename... Ts>
void f(Ts&&... args)                    //函数的参数包
{
    auto myLambda = [&...args = std::forward<Ts>(args)](){};
```

}
```

另外，C++20 增加了废弃 lambda 表达式隐式捕获 this 指针（deprecate implicit capture of this）的做法，现在不推荐使用"[=]"在 lambda 表达式中隐式捕获 this 指针，首选"[=, this]"或"[=, *this]"在 lambda 表达式中显式捕获 this 指针。例如：

```
struct int_value{
    int n = 0;
    auto getter_fn() {
        return [=]() {return n;};              //不推荐
        return [=, *this]() {return n;};       //推荐
    }
};
```

2.5 本章小结

本章详细介绍了 C++11 新增的 lambda 表达式，以及后续 C++版本对 lambda 表达式的扩展。在内容安排上采用 C++11/14/17/20 的顺序。

lambda 表达式定义了一个匿名函数（lambda 函数），该函数可随处定义随处使用，不具有代码重用的性质，可以使代码更加灵活简洁。lambda 函数与普通函数类似，也有参数列表、返回类型和函数体，只是它的定义方式更为简洁，并且可以在函数内部定义。

编译器会把 lambda 表达式翻译成一个匿名类或者匿名函数，并重载 operator()。捕获列表将成为匿名函数中的构造函数参数。如果参数是通过值捕获方式来捕获的，则会在匿名函数中创建相应类型的数据成员。如果参数是通过引用捕获方式来捕获的，则会在匿名函数中创建相应类型的引用数据成员。lambda 函数参数列表中声明的变量，该参数将成为调用运算符 operator() 的参数。

lambda 函数的捕获列表可以通过值捕获或引用捕获的方式来捕获上下文变量，并在 body 中直接使用。如果在一个类的成员函数中使用 lambda 表达式，则可以通过引用捕获方式来捕获该类的成员变量。lambda 表达式既可以保存 auto 类型的变量，也可以把 auto 类型的变量赋值给 std::function 对象。

lambda 表达式通常是作为 STL 算法库<algorithm>中函数的最后一个参数使用的，如 std::sort、std::for_each 和 std::find_if 等。lambda 表达式还可以用于回调函数（参见第 6 章）。

随着 C++后续版本的推出，lambda 表达式也将不断得到扩展，一些原有的特性可能被废弃，一些新的特性可能会被引入，因此需要注意编译器所支持的 C++版本。
```

第 3 章
面向对象的技术

面向对象的技术是 C++的核心。我们初次接触 C++时会了解一些基本概念，如封装、继承、重载、多态性、构造函数、析构函数、成员函数、成员变量、虚函数、静态变量等编程语言概念。由于篇幅限制，我们不打算在本章里对上面的概念一一介绍，而是重点介绍自 C++11 版本起引进的新的面向对象技术，主要涉及以下的内容：

- 转换构造函数（converting constructor）和显式构造函数（explicit constructor）。
- 拷贝构造函数（copy constructor）。
- 拷贝赋值运算符（copy assignment operator）。
- 移动构造函数（move constructor）。
- 移动赋值运算符（move assignment operator）。
- 委派构造函数（constructor delegation）。
- 继承构造函数（constructor inheritance）。
- 显式默认（default）函数和显式删除（deleted）函数。
- 私有构造函数（private constructor）。
- 虚函数表（VTable）和虚函数指针（__vptr）。
- 显式重写（override）和终止重写（final）。
- 常量表达式虚函数（constexpr virtual function）。
- 重载赋值运算符（overloading operator）。
- 函数调用运算符的重载（overloading of function-call operator）。
- 多重继承（multiple inheritance）。
- 多级继承（multilevel inheritance）。
- 虚继承（virtual inheritance）。
- 左值引用（lvalue reference）和右值引用（rvalue reference）。
- 移动语义（move semantic）。
- 引用限定符（ref-qualifier）。

本章分 5 节来介绍上述的面向对象技术。

3.1 构造函数与析构函数

类和结构（以下统称为类）的构造函数是一种特殊方法，在创建类对象时会自动调用相应的构造函数。构造函数通常用于初始化新对象的数据成员，这就是它为什么被称为构造函数的原因。

构造函数名与类同名。根据构造函数在何种情况下使用,可将构造函数分为以下 4 类:
- 默认构造函数(default constructor)。
- 参数构造函数(parameter constructor)。
- 拷贝构造函数(copy constructor)。
- 移动构造函数(move constructor)。

3.1.1 转换构造函数和显式构造函数

可以使用一个实参调用构造函数来执行隐式转换的构造函数称为转换构造函数,所有的构造函数都默认是转换构造函数。我们前面说过,构造函数的任务是初始化成员变量。初始化可分为拷贝初始化和直接初始化。我们先看看转换构造函数的示例:

【例 3-1】

```cpp
#include <iostream>
class foo{
public:
    foo(int x): m_x{x} {}              //转换构造函数
    int getX() const { return m_x; }
private:
    int m_x;
};

void printFoo(foo f)                   //foo 是参数类型
{
    std::cout << f.getX();
}

int main(){
    printFoo(5);                       //传入整数 5 给 printFoo,但它被隐式转化成 foo 对象
    return 0;
}
```

在上面的示例中,编译器将找到转换构造函数 foo(int),该函数将 int 的值 5 转换为 foo 对象。

【例 3-2】

```cpp
struct A{
    A() {}                 //转换构造函数
    A(int) {}              //转换构造函数
    A(int, int) {}         //转换构造函数
};
int main(){
    A a1 = 1;              //正确:通过构造函数 A::A(int)进行拷贝初始化
    A a2(2);               //正确:直接初始化,选择 A::A(int)
    A a3{4, 5};            //正确:直接列表初始化,选择 A::A(int, int)
    A a4 = {4, 5};         //正确:拷贝初始化列表,选择 A::A(int, int)
    A a5 = (A)1;           //正确:显式静态类型转换,直接初始化
    return 0;
}
```

如果在创建类的对象时提供了不正确数据类型的实参,则编译器不能成功地进行隐式类型转换,将报告编译错误。

相对于隐式类型转换的参数构造函数,显式构造函数明确指定构造函数是显式的,即不能用于隐式转换、拷贝初始化以及拷贝初始化列表。显式构造函数在构造函数前面加上 explicit。例如:

【例 3-3】

```
struct B{
    explicit B() { }
    explicit B(int) { }
    explicit B(int, int) { }
};

int main(){
    B b1 = 1;           //错误:拷贝初始化不能使用 B::B(int)
    B b2(2);            //正确:直接初始化,使用 B::B(int)
    B b3{4, 5};         //正确:直接初始化,使用 B::B(int, int)
    B b4 = {4, 5};      //错误:拷贝初始化列表不能使用 B::B(int, int)
    B b5 = (B)1;        //正确:显式静态转换,直接初始化
    B b6;               //正确:默认初始化
    B b7{};             //正确:直接列表初始化
    B b8 = {};          //错误:拷贝初始化列表使用显式构造函数
    return 0;
}
```

现在流行的做法是使任何带有默认值的单个参数的构造函数显式化。同样,对于包括具有多个参数的构造函数,如果其中大多数或全部参数都具有默认值,则也使用显式构造函数。这将禁止编译器使用该构造函数进行隐式转换。如果确实需要隐式转换,则仅考虑非显式构造函数。但如果编译器找不到非显式构造函数来执行转换,将报告编译错误。归纳起来:

- 拷贝构造函数和移动构造函数不应该显式化。
- 不带参数的默认构造函数一般不显式化。
- 带多个参数且具有默认值的构造函数一般显式化。
- 带单个参数的构造函数一般显式化。

3.1.2 拷贝构造函数

拷贝构造函数使用类的现有对象创建一个新对象,并且对新对象的成员进行拷贝初始化。如果不显式定义拷贝构造函数,则编译器会隐式生成类的拷贝构造函数。

需要牢记的是:拷贝构造函数将对同一类对象的引用作为参数而不采用传值方式。如果在拷贝构造函数中按值传递对象,则会导致对拷贝构造函数本身的递归调用。发生这种情况是因为传递值涉及创建副本,而创建副本涉及调用拷贝构造函数,从而导致无限循环。使用引用可以避免这种递归调用。因此,我们使用对象的引用来避免无限循环。

什么时候需要定义拷贝构造函数呢?当对象拥有指针或不可共享的引用(如对文件的引用)时,通常需要用户显式定义拷贝构造函数。在这种情况下,还需要编写析构函数和重载赋值运算符。如果不定义自己的拷贝构造函数,则编译器会为每个类创建一个默认的拷贝构造函数,但该拷贝构造函数在对象之间执行成员级拷贝。编译器创建的拷贝构造函数通常工作正常,但当对象

具有指针或任何在运行时分配资源（如文件句柄、网络连接等）时，编译器创建的拷贝构造函数是无法胜任的，因为它在对象之间执行浅拷贝。请看下面的例子。

【例 3-4】

```cpp
#include <cstring>
#include <iostream>
using namespace std;

class MyString {
private:
    char* s;
    int size;
public:
    MyString(const char* str = NULL);
    ~MyString() { delete[] s; }
    MyString(const MyString&);          //拷贝构造函数
    void print(){   cout << s << endl;}
};

MyString::MyString(const char* str){
    size = strlen(str);
    s = new char[size + 1];
    strcpy(s, str);
}

MyString::MyString(const MyString& old_str){
    size = old_str.size;
    s = new char[size + 1];             //必须分配内存，用于存放字符串
    strcpy(s, old_str.s);               //拷贝字符串
}

int main(){
    MyString str0("Hello, world");
    MyString str1 = str0;               //注意：这里调用拷贝构造函数
    str0.print();
    str1.print();
    return 0;
}
```

3.1.3 拷贝赋值运算符

拷贝赋值运算符和拷贝构造函数类似，都可用于初始化一个对象。但是，它们之间存在一些基本区别：拷贝赋值运算符用于已经创建的对象的初始化，即从一个现有的对象分配新值给另一个已经创建的对象，它的语法格式如下：

```cpp
className& operator=(const className& t){
    //....
    return *this;
}
```

需要牢记的是：拷贝赋值运算符将对同一类的对象的引用作为参数而采用传值方式。其中的原因和拷贝构造函数类似。如果不定义拷贝赋值运算符，则编译器会隐式生成类的拷贝赋值运算符。

什么时候需要定义拷贝赋值运算符呢？简单地说，当定义显式拷贝构造函数时，也需要显式定义拷贝赋值运算符。现在给上面的类 MyString 加上拷贝赋值运算符。

```cpp
MyString& MyString::operator=(const MyString& old_str){
    if(*this != old_str){
        if(s != NULL)
            delete []s;                    //记住：这里必须先删除现有的内存资源
        size = old_str.size;
        s = new char[size + 1];            //必须分配内存，用于存放字符串
        strcpy(s, old_str.s);              //拷贝字符串
    }
    return *this;
}
```

3.1.4 移动构造函数

拷贝构造函数使用左值引用和复制语义（即将现有对象的实际数据复制到另一个对象，而不是使另一个对象指向堆中的现有对象）。移动构造函数使用右值引用和移动语义（即指向内存中已经存在的对象）。当构造函数将右值引用作为参数时，称为移动构造函数。

移动构造函数的语法格式如下：

```cpp
className (className&& obj) {
    //移动构造函数的主体
}
```

移动构造函数可移动堆中的资源，使声明对象的指针指向临时对象的数据，并清空临时对象的指针，因此，移动构造函数可以防止不必要地拷贝内存中的数据。移动构造函数和移动赋值运算符都涉及移动语义，我们在下面的移动语义部分做进一步的介绍。

【例 3-5】

```cpp
#include <iomanip>
#include <iostream>
#include <string>
#include <utility>

struct A{
    std::string s;
    int k;

    A():s("test"), k(-1) {}
    A(const A& a) : s(a.s), k(a.k) { std::cout << "move failed!\n";}
    A(A&& a) noexcept :
        s(std::move(a.s)),              //显式移动类的成员变量
        k(std::exchange(a.k, 0))        //显式移动非类的成员变量
    {}
```

```
    };

    int main(){
        A a1;
        A a2 = std::move(a1);        //a1 已经创建于内存中，从 a1 移动构造 a2, a1 的 s 内容被清空。
        std::cout << "After move, a1.s = " << std::quoted(a1.s)
            << " a1.k = " << a1.k << '\n';
        return 0;
    }
```

3.1.5 移动赋值运算符

C++11 添加了移动赋值运算符，以进一步加强 C++中的移动语义（move semantic）。移动赋值运算符类似于拷贝赋值运算符，但不是拷贝数据，而是将给定数据的所有权移动到目标对象，无须进行任何额外的拷贝，源对象将保持有效但不确定的状态。移动赋值运算符的语法格式如下：

```
MyClass& operator= (MyClass&& other) noexcept {
    //...
    return *this;
}
```

注意到，移动赋值运算符使用特殊的&&引用限定符，它表示右值引用。

【例 3-6】

```
#include <iostream>
#include <string>
#include <utility>

class A{
public:
    std::string s;
    A() : s("test") {}
    A(const A& o) : s(o.s) { std::cout << "move failed!\n"; }
    A(A&& o) : s(std::move(o.s)) {}
    A& operator=(const A& other){
        s = other.s;
        std::cout << "copy assigned\n";
        return *this;
    }
    A& operator=(A&& other){
        s = std::move(other.s);
        std::cout << "move assigned\n";
        return *this;
    }
};
A f(A a) { return a; }

int main(){
    A a1;
```

```
        std::cout << "Trying to move-assign A from rvalue temporary\n";
        a1 = f(A());                        //从 f()返回的临时变量进行移动赋值
        std::cout << "Trying to move-assign A from xvalue\n";
        return 0;
}
```

3.1.6 委派构造函数

C++11 引入的委派构造函数能够调用同一类的另一个构造函数，例如：

```
class A{
public:
    A() {}
    //成员初始化，无委派构造函数
    A(string s) : str{s} {}
    //委派构造函数
    A(string s, double d):A(s) {
        a = d;
    }
    //委派构造函数不能再进行初始化，只能使用赋值语句
    A(string s, double d):A(s), a{d} {}          //错误

private:
    double a {1.0};
    string str;
};
```

委派构造函数会导致构造函数的递归调用（constructor1 调用 constructor2，后者又调用前者），并且在堆栈溢出前不会引发任何错误，因此需要避免出现递归调用。

```
class B{
public:
    //不能出现如下的递归调用
    B() : B(6, 3){}
    B(int my_max, int my_min) : B(){}
    private:
    int max;
    int min;
};
```

3.1.7 继承构造函数

C++11 引入了一种构造函数继承形式，我们可以通过 using-declaration 指示编译器生成一组构造函数，这些构造函数采用与基类构造函数相同的参数，并将这些参数转发到基类。虽然这种转发被正式称为继承，但现在这些参数只是由编译器生成的，而不是我们显式编写的。

【例 3-7】

```
#include <iostream>
using namespace std;
```

```cpp
struct Base {
    Base(int a) : i(a) {}
    int i;
};
struct Derived : Base {
    Derived(int a, std::string s) : Base(a), m(s) {}    //显式定义的构造函数
    using Base::Base;          //继承了基类的构造函数，等价于 Derived(int a) : Base(a), m() {}
    std::string m;
};
int main() {
    Derived d(5);
    return 0;
}
```

在本例中，Derived 有两个构造函数（不包括拷贝构造函数和移动构造函数），第一个是显示定义的构造函数，它有两个参数，一个参数用于接收一个 int 类型变量，另一个参数用于接收一个字符串；第二个构造函数是使用 using Base::Base 继承了基类的构造函数，等价于

```cpp
Derived(int a) : Base(a), m() {}
```

3.1.8 显式默认函数和显式删除函数

显式默认函数声明是 C++11 引入的一种新的函数声明形式，它允许将说明符"=default"追加到函数声明的末尾，将该函数声明为显式默认函数，从而使编译器为显式默认函数生成默认实现，这比手动编写的函数更有效，因为无须再编写函数体了。

我们可以使用默认说明符来创建无参数的显式默认函数，例如：

```cpp
struct A {
    A() = default;
    A(int x) : x{x} {}
    int x {1};
};
A a;                    //a.x == 1
A a2 {123};             //a2.x == 123
```

下面看看有继承的情况：

```cpp
struct B {
    B() : x{1} {}
    int x;
};
struct C : B {
    C() = default;
};
C c;                    //调用 B::B
                        //c.x == 1
```

显式默认函数必须是特殊的成员函数（默认构造函数、拷贝构造函数、移动构造函数、析构函数等）。例如，以下代码给出了不能使用默认标识符的非特殊成员函数：

```cpp
class B {
```

```cpp
public:
    int func() = default;              //错误：func()不是特殊成员函数
    B(int, int) = default;             //错误：B(int, int)不是特殊成员函数
    B(int = 0) = default;              //错误：B(int=0)有默认参数
};
```

C++11 引入了"=delete"运算符，通过将该运算符添加到函数声明的末尾，可禁止使用该成员函数。使用"=delete"说明符的任何成员函数都称为显式删除函数。如果我们不希望在类的外面创建多个对象，则可以把拷贝构造函数和重载赋值运算符禁止，即加上"=delete"。例如：

```cpp
class A {
public:
    A(int x): m(x){}
    A(const A&) = delete;              //删除拷贝构造函数
    A& operator=(const A&) = delete;   //删除重载赋值运算符
    int m;
};
A x {123};
A y = x;                               //错误：调用已经被删除了的拷贝构造函数
y = x;                                 //错误：operator 已经被删除
```

3.1.9 私有构造函数

在某些情况下，我们可能需要定义私有构造函数，以阻止在类的外面创建类的对象实例。在软件设计中，如果使用唯一的静态全局类对象，则需要定义私有构造函数。单例模式便是一个很好的例子，它的示例代码如下：

【例 3-8】

```cpp
#include <iostream>
class Singleton {
public:
    //定义一个静态函数来访问 Singleton 实例
    static Singleton& getInstance(){
        if (!instance) {
            instance = new Singleton();
        }
        return *instance;
    }
    //删除拷贝构造函数和重载赋值运算符
    Singleton(const Singleton&) = delete;
    Singleton& operator=(const Singleton&) = delete;

private:
    //私有构造函数阻止在类的外部创建对象实例
    Singleton(){
        std::cout << "Singleton instance created." << std::endl;
    }
    //私有析构函数阻止在类的外部删除对象实例
    ~Singleton(){
```

```
                std::cout << "Singleton instance destroyed." << std::endl;
                delete instance;
            }
            static Singleton* instance;
    };
    Singleton* Singleton::instance = nullptr;
    int main(){
        Singleton& singleton = Singleton::getInstance();
        //试图调用构造函数来创建 Singleton 的其他实例，将报告编译错误
        return 0;
    }
```

在单例模式中，我们不仅要将构造函数和析构函数放在 private 部分，还废弃了拷贝构造函数和移动构造函数。

3.2 虚函数与多态性

虚函数是 C++的重要概念，也是 C++程序设计的重要技术。虚函数是在基类中使用 virtual 关键字声明并在派生类中重新定义（重写）的成员函数。定义虚函数的目的是实现运行时的多态性，即在运行时根据对象的类型确定要调用对象的函数版本。

虚函数有如下规则：
- 虚函数不能是静态的。
- 虚函数可以是另一个类的友元函数（friend function）。
- 只能通过基类类型的指针或引用来访问虚函数，以实现运行时的多态性。
- 虚函数的原型在基类和派生类中必须是完全相同的。
- 虚函数始终在基类中定义，并在派生类中重写，但在派生类中不需要在函数前面加 virtual。派生类不强制覆盖虚函数，在这种情况下，可使用函数的基类版本。
- 类中有虚成员函数时，说明类会被继承，所以析构函数应该定义为虚函数。
- 基类中有指针变量时，析构函数必须定义为虚函数。
- 如果基类中的虚成员函数后面是"= 0;"，则该虚函数为纯虚函数，不能直接生成基类的指针或引用对象。
- 如果基类中的成员函数全部是纯虚函数，则该基类一般称为抽象类，在派生类中一定要实现纯虚函数。

虚函数运行时的多态性只能通过基类的指针或引用来实现。基类指针可以在指向基类的对象同时也指向派生类的对象。实际上，指向基类对象的指针包含派生类对象的地址。迟绑定（运行时绑定）是根据指针的内容（即指针指向的位置）完成绑定的，早绑定（编译时绑定）是根据指针的类型完成绑定的。在 C++中，虚函数采用迟绑定，非虚函数采用早绑定，从而呈现多态性。我们先看看下面的例子。

【例 3-9】

```
#include <iostream>
using namespace std;
class base {
```

```cpp
public:
    void f1() { cout << "base-1\n"; }
    virtual void f2() { cout << "base-2\n"; }        //虚函数
    virtual void f3() { cout << "base-3\n"; }        //虚函数
    virtual void f4() { cout << "base-4\n"; }        //虚函数
};
class derived : public base {
public:
    void f1() { cout << "derived-1\n"; }
    void f2() { cout << "derived-2\n"; }             //虚函数
    void f4(int x) {cout << "derived-4\n";}
};
int main(){
    base* p;                                         //指向基类的指针
    p = new derived;
    p->f1();        //早绑定，base 中的 f1()，因为 f1()不是虚函数
    p->f2();        //迟绑定，绑定 derived 中的 f2()
    p->f3();        //迟绑定，绑定 base 中的 f3()
    p->f4();        //迟绑定，绑定 base 中的 f4()
    p->f4(5);       //非法调用，f4(int x)不是 base 中的虚函数
    return 0;
}
```

我们接下来看看纯虚函数和抽象类的示例，需要记住抽象类是不能实例化的。

【例 3-10】

```cpp
class Abstract{                  //抽象类
public:
    virtual void f() = 0;        //纯虚函数
};

class Concrete : public Abstract{
public:
    void f() override {}         //非纯虚函数
    virtual void g();            //非纯虚函数
};

class Abstract1 : public Concrete{
public:
    void g() override = 0;       //纯虚函数，Abstract1 为抽象类
};

int main(){
    Abstract a;                  //错误：Abstract 是抽象类，不能实例化
    Concrete b;                  //正确
    Abstract& a = b;             //正确
    a.f();                       //调用 Concrete::f()
    Abstract1 a2;                //错误：Abstract1 不能实例化
    return 0;
}
```

3.2.1 虚函数表和虚函数指针

我们来进一步探究 C++中虚函数的多态性。在编译时,每个具有虚函数的类都将获得一个虚函数表(virtual table)。虚函数表是一个仅包含虚函数内存地址的表,并且是按类维护的。虚函数表是由编译器在编译时创建的。

当我们创建具有虚函数的对象时,编译器会向类的对象中添加一个名为__vptr(virtual table pointer)的隐藏成员,该成员是指向虚函数表的指针,并且按对象实例进行维护,即每个对象都维护一个虚函数表。__vptr 可被继承到所有的派生类。如果从基类创建多个派生类,则所有派生类中的__vptr 都指向特定类的同一虚函数表。虚函数表和虚函数指针如图 3.2-1 所示。

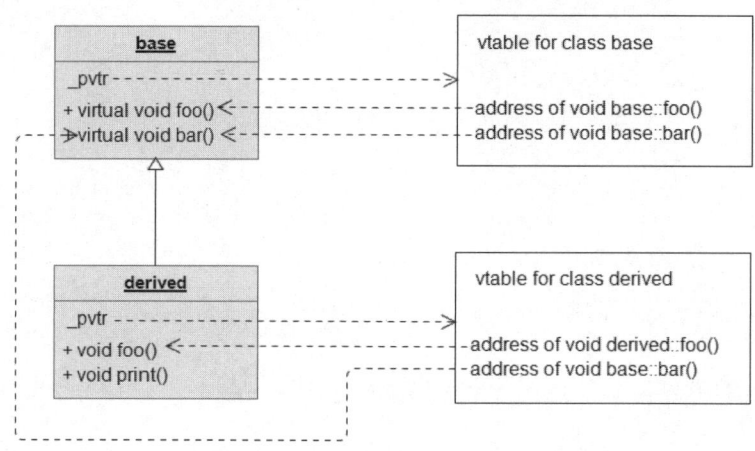

图 3.2-1　虚函数表和虚函数指针示意

虚函数在运行时通过查询虚函数表决定调用相应的函数版本。

3.2.2 显式重写和终止重写

修饰符 override 可用于指定一个派生类的虚函数覆盖其父类中的相同签名的虚函数。如果虚函数未覆盖父函数的相同签名虚函数,则会引发编译器错误。相同签名是指函数名称、参数、返回类型完全相同。

```
class A {
public:
    virtual void foo();
    void bar();
};
class B : public A {
public:
    void foo() override;      //正确: B::foo overrides A::foo
    void bar() override;      //错误: A::bar 不是虚函数
    void baz() override;      //错误: A::baz 没有声明
};
```

当修饰符 final 修饰一个虚函数时,表示不能在派生类中重写该虚函数。例如:

```cpp
class A {
public:
    virtual void foo();                //虚函数
};
class B : public A {
public:
    virtual void foo() final;          //虚函数 foo 的最后版本,不能被重写
};
class C : public B {
public:
    virtual void foo();                //错误:foo 在 B 中被标记为 final
};
```

当修饰符 final 用于类时,表示该类不能被继承。例如:

```cpp
class A final {};
class B : public A {};                 //错误:基类 A 被标记为 final
```

3.2.3 常量表达式虚函数

关键字 constexpr 和 virtual 可以在一起工作吗?virtual 表示运行时的多态性,而 constexpr 表示常量表达式。看起来它们似乎不相干或者还是矛盾的。

我们首先来看看下面的代码。我们需要处理一个产品列表,并且要检查每一种产品是否适合给定的包装盒尺寸。整个代码似乎看不出什么问题。但如果我们希望在 C++17 对应的环境中编译执行这样的代码,则会出问题,主要原因是 virtual 运行时的多态性。在 C++17 中,我们必须将运行时的多态性替换为静态多态性。

【例 3-11】

```cpp
#include <cassert>
class Box {
public:
    double width{0.0};
    double height{0.0};
    double length{0.0};
};
class Product {
public:
    virtual ~Product() = default;
    virtual Box getBox() const noexcept = 0;              //纯虚函数
};
class Notebook : public Product {
public:
    Box getBox() const noexcept override {
        return {.width = 30.0, .height = 2.0, .length = 30.0};
    }
};
class Flower : public Product {
public:
```

```cpp
        Box getBox() const noexcept override {
            return {.width = 10.0, .height = 20.0, .length = 10.0};
        }
    };
    bool canFit(const Product &prod, const Box &minBox) {
        const auto box = prod.getBox();
        return box.width < minBox.width && box.height < minBox.height && box.length < minBox.length;
    }
    int main() {
        Notebook nb;
        Box minBox{100.0, 100.0, 100.0};
        assert(canFit(nb, minBox));
        return 0;
    }
```

我们在虚函数前面加上 constexpr，问题就解决了。加上 constexpr 的主要作用是可以轻松地将现有代码转换为编译时版本，相当于去除了多态性，实现静态编译。当代码处于编译级别时，必须预先知道所有类型。加上 constexpr 就可以显式地告诉编译器执行去虚拟化，这样就不会生成任何多态性代码。

【例 3-12】

```cpp
#include <cassert>
struct Box {
    double width{0.0};
    double height{0.0};
    double length{0.0};
};
struct Product {
    constexpr virtual ~Product() = default;
    constexpr virtual Box getBox() const noexcept = 0;
};
struct Notebook : public Product {
    constexpr Box getBox() const noexcept override {
        return {.width = 30.0, .height = 2.0, .length = 30.0};
    }
};
struct Flower : public Product {
    constexpr Box getBox() const noexcept override {
        return {.width = 10.0, .height = 20.0, .length = 10.0};
    }
};
constexpr bool canFit(const Product &prod, const Box &minBox) {
    const auto box = prod.getBox();
    return box.width < minBox.width && box.height < minBox.height && box.length < minBox.length;
}
int main() {
    constexpr Notebook nb;
    constexpr Box minBox{100.0, 100.0, 100.0};
    static_assert(canFit(nb, minBox));
```

```
        return 0;
}
```

3.3 重载

3.3.1 重载赋值运算符

在 C++中，重载是重要的功能之一。重载包括函数重载和运算符重载。函数重载是面向对象编程的一个重要特性，其中两个或多个函数可以具有相同的名称但不同的参数。由于不同的任务而重载相同名称的函数，称为函数重载。在函数重载中，函数名称应相同，参数应不同。函数重载被视为 C++中多态性特征的一个示例。运算符重载体现了编译时多态性，它是一种在不改变原始含义的情况下为 C++中的现有运算符赋予特殊含义的方法。

【例 3-13】

```cpp
#include <iostream>
using namespace std;

class Complex {
private:
    int real, imag;
public:
    Complex(int r, int i){
        real = r;
        imag = i;
    }
    //重载复数加法运算符
    Complex operator+(Complex const& obj){
        Complex ans;
        ans.real = real + obj.real;
        ans.imag = imag + obj.imag;
        return ans;
    }
    //重载复数乘法运算符
    Complex operator*(Complex const& obj){
        Complex ans;
        ans.real = (real * obj.real) - (imag * obj.imag);
        ans.imag = (imag + obj.real) + (real * obj.imag);
        return ans;
    }
    void print() {cout<< real <<" + "<< imag << "i" << '\n';}
};
int main(){
    Complex c1(8, 5), c2(6, 4);
    Complex c3 = c1 + c2;
    Complex c4 = c1 * c2;
    c3.print();
```

```
        c4.print();
        return 0;
}
```

3.3.2 函数调用运算符的重载

函数调用运算符是 operator()，用于函数调用和参数传递。函数调用运算符由称为函数对象的类的实例重载，当重载函数调用运算符时，需要创建一个可用于传递参数的运算符函数。

在面向对象编程语言中，operator()可以被视为普通的运算符，类的对象可以调用 operator()，就像调用其他重载赋值运算符一样。我们先看看下面的示例。

【例 3-14】

```cpp
#include <iostream>
using namespace std;
#define N 3
#define M 3

class Matrix {
private:
    int arr[N][M];
public:
    //声明重载赋值运算符 ">>" 为友元函数
    friend istream& operator >>(istream&, Matrix&);
    //声明重载赋值运算符 "<<" 为友元函数
    friend ostream& operator<<(ostream&, Matrix&);
    //声明函数调用运算符
    int& operator()(int, int);
};
//重载赋值运算符 ">>"，友元函数可以访问类的私有成员
istream& operator >>(istream& cin, Matrix& m) {
    int x;
    for (int i = 0; i < N; i++) {
        for (int j = 0; j < M; j++) {
            cin >> m(i, j);    //m(i, j)为函数调用运算符
        }
    }
    return cin;
}
//重载赋值运算符 "<<"，友元函数可以访问类的私有成员
ostream& operator <<(ostream& cout, Matrix& m) {
    for (int i = 0; i < N; i++) {
        for (int j = 0; j < M; j++) {
            cout << m(i, j) << " ";           //m(i, j)为函数调用运算符
        }
        cout << endl;
    }
    return cout;
}
```

```
//重载的函数调用运算符
int& Matrix::operator()(int i, int j) {
    return arr[i][j];
}
int main() {
    Matrix m;
    cin >> m;
    cout << m;
    return 0;
}
```

3.4 继承

3.4.1 多重继承

多重继承是 C++的一个特性，一个类可以从多个父类继承。当调用类的构造函数生成类的实例时，被多重继承的父类构造函数被触发调用，并且调用顺序与继承它们的顺序相同。

【例 3-15】

```
#include <iostream>
using namespace std;

class Mammal{
public:
    Mammal(){
        cout << "Mammals can give direct birth." << endl;
    }
};
class Bird{
public:
    Bird(){
        cout << " Birds can fly." << endl;
    }
};
class Bat: public Mammal, public Bird {};
int main() {
    Bat b;
    return 0;
}
```

多重继承带来的最明显的问题是函数重载引起的歧义。假设两个基类具有相同的函数，且该函数在派生类中未被重写，如果派生类对象尝试调用该函数，则编译器将显示错误。这是因为编译器不知道该调用哪个基类的函数。例如：

```
class base1 {
public:
    void func( ) {....}
```

```
};
class base2 {
    void func( ) {....}
};
class derived : public base1, public base2 {};

int main() {
    derived d;
    d.func()                //错误：因为编译器不知道该调用哪个基类的函数
    d.base1::func();        //正确
    d.base2::func();        //正确
    return 0;
}
```

3.4.2 多级继承

在C++中，多级继承是指类的继承关系中不仅有父子关系，还有祖孙关系，即派生类在继承一个类时，它又会被另一个类继承。多级继承有时会出现"钻石"问题，即当一个类的两个父类具有共同的基类时，就会发生"钻石"问题。例如，在图3.4-1中，TA类获取Person类的所有属性的两个副本，这会导致歧义。

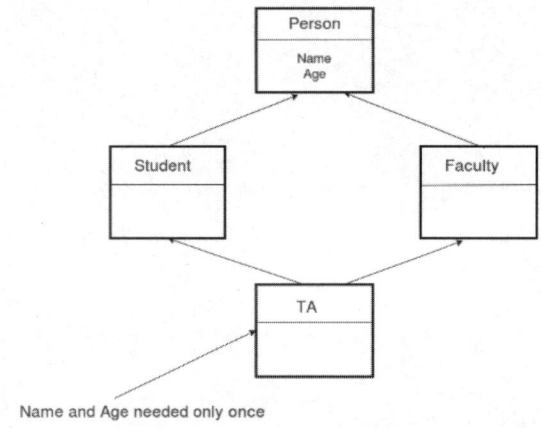

图3.4-1 "钻石"问题示例

【例3-16】

```
#include<iostream>
using namespace std;

class Person {
public:
    Person(int n, string s), age(n), name(s){}
    int age;
    string name;
};
class Faculty : public Person{
public:
```

```cpp
        Faculty(int n, string s):Person(n,s){}
};
class Student : public Person {
public:
        Student(int n, string s):Person(n,s){}
};
class TA : public Faculty, public Student {
public:
        TA(int n, string s):Student(n, s), Faculty(n, s) {}
};
int main() {
        TA ta(30, "Alex");
        return 0;
}
```

在上面的代码中,当创建类 TA 的对象 ta 时,需要调用类 Student 和类 Faculty 的构造函数,它们又都调用类 Person 的构造函数,导致类 Person 被构造两次,因此 ta 中有两份类 Person 中的成员 age 和 name,将引起歧义。解决这个问题的方法是使用虚继承。

3.4.3 虚继承

虚继承是一种 C++技术,可确保孙辈派生类仅继承祖父辈基类成员变量的一个副本。在下面的例子中,通过虚继承,可以使派生类的 Bat 对象只保留基类 Animal 的成员变量 species_name 的一个副本。

【例 3-17】

```cpp
#include <iostream>
using namespace std;

class Animal {
private:
        string species_name;
public:
        Animal(const string& name) : species_name(name){
                cout << "Animal constructor called" << endl;
        }
        void show_species() const {
                cout << "This animal belongs to the species: " << species_name << endl;
        }
};

class Bird : virtual public Animal {
public:
        Bird(const string& name) : Animal(name){
                cout << "Bird constructor called" << endl;
        }
};
```

```cpp
class Mammal : virtual public Animal {
public:
    Mammal(const string& name) : Animal(name) {
        cout << "Mammals constructor called" << endl;
    }
};

class Bat : public Bird, public Mammal {
public:
    //使用虚继承后，这里直接初始化基类 Animal
    Bat(): Animal("Bat"), Bird("Bat"), Mammal("Bat") {}
    void show_info() {
        cout << "It's a unique animal with some details:" << endl;
        show_species();
    }
};

int main() {
    Bat my_bat;
    my_bat.show_info();
    return 0;
}
```

3.5 其他杂项

3.5.1 左值引用和右值引用

左值是仅出现在赋值符号左边的变量或对象，有名称和程序可访问的地址。变量的名字、函数的名字（注意不是函数调用）、类成员的名字都是左值。

左值引用是使用符号&对左值变量或对象进行引用，如 int& ref 表示 ref 可以引用一个整型变量。

C++11 新增了一条规则：引用可以绑定到左值，但对常量的引用可以绑定到右值。具体来说：对某个数据类型 T 的引用只能绑定到类型 T 的左值上，例如下面的代码：

```
int a;
int &r = a;              //正确
int& ref = 2050;         //错误：因为 2050 是常量，是右值
```

在引用类型 T 的常量时不一定非得绑定到类型 T 的左值上，只要表达式产生的值可以转换成类型 T 即可。例如下面两句代码都是正确的。

```
const int& ref = 2050;
const double& ref = 2050;
```

因为右值 2050 这个表达式产生的值可以转换为 int 或 double，因此 C++编译器就会首先创建了一个临时对象（第一句代码的临时对象是 int 类型的，第二句代码的临时对象是 double 类型的），然后将 ref 引用绑定到临时对象上。

在 C++11 中引进了右值引用。右值是指赋值符号右边的常量、表达式。被右值引用的常量或者对象只有一个内存地址（临时对象），没有对象名称。右值引用使用两个&&。例如：

```
int && ref = 2060;              //正确：右值引用绑定到常量
int a = 10;
int && ref = 2*a+10;            //正确：右值引用可以绑定到表达式
int b = 7030;
int&& ref = b;                  //错误：右值引用不能绑定左值变量
const int&& ref = b ;           //错误：不能绑定右值引用到左值变量
```

3.5.2 移动语义

我们现在来讨论一下移动语义。移动语义和复制语义相对应。复制语义是指将源对象的成员值复制到目标对象中，生成两个具有相同状态的独立对象。然而，许多现实生活中的实体并不是这样的，你可以移动它们，但不能复制它们或者不应该复制它们。例如，当搬家时，您最终只是在搬迁后拥有一个新家，而不是拥有两个相同的家。本节将介绍移动语义，并讨论 C++中复制对象和移动对象的不同应用。

在许多编程任务中，资源仅从一个对象移动到另一个对象，从而清空移动过程中的源对象。这些移动对象的语义和形式属性是一种新的 C++范式，可以提高代码效率并更准确地模拟现实世界的情况。

文本编辑器和图形用户界面（GUI）编辑系统允许我们选择对象并将其剪切或复制到新的位置。剪切将原始对象移动到新的位置，而复制会在新位置创建一个新对象，将原始对象保留在其原始位置。这是移动操作和复制操作的经典应用。

移动语义的话题有点意思，为什么要让编程语言支持移动语义呢？因为复制对象是一项成本高昂的操作，它需要消耗大量内存和 CPU 资源，尤其是在涉及图像、视频剪辑或大量数据文件等操作时。您可能会惊讶地发现，在许多情况下，C++是以复制对象为默认方式的。例如，函数以值的方式返回一个对象。

移动构造函数和拷贝构造函数是相互对应的。移动构造函数不是将源对象复制到目标对象中，而是"窃取"源对象的资源，将它们移动到目标对象中。因此，在调用移动构造函数之后，源对象将保持未指定状态，而目标对象将成为这些资源的独占者。移动构造函数具有以下签名：

```
C::C(C&&);
```

移动赋值运算符和拷贝赋值运算符是相互对应的，它"窃取"源对象的资源，将它们转移到*this。移动赋值运算符的签名格式如下：

```
C& C::operator=(C&&);
```

移动构造函数和移动赋值运算符统称为移动特殊成员函数。这产生了一个两难境地：一个类应该定义两者中的哪一个（如果有的话）。

在某些情况下，C++在未显式声明特殊成员函数的类中隐式地声明它们，如拷贝构造函数、拷贝赋值运算符等。在 C++11 中，该规则也适用于两个移动特殊成员函数，因此如果一个类的定义中未显式声明移动赋值运算符，则当满足以下所有的条件时，才会隐式地声明为默认运算符：

● 类中没有用户声明的拷贝构造函数。
● 类中没有用户声明的移动构造函数。

- 类中没有用户声明的拷贝赋值运算符。
- 类中没有用户声明的析构函数。
- 移动赋值运算符不会被隐式定义为已删除。

下面我们给出一个完整的类示例代码,展示拷贝构造函数、移动构造函数、拷贝赋值运算符和移动赋值运算符的用法。

【例 3-18】

```cpp
#include <iostream>
#include <algorithm>
#include <vector>

class MyString{
public:
    explicit MyString(const char * str){
        _length = strlen(str);
        _data = new char[_length+1];
        strcpy(_data, str);
    }
    ~MyString(){
        if (_data != nullptr){
            delete[] _data;
        }
    }
    //拷贝构造函数
    MyString(const MyString& other)
    : _length(other._length), _data(new char[other._length]){
        std::copy(other._data, other._data + _length, _data);
    }
    //拷贝赋值运算符
    MyString& operator=(const MemoryBlock& other{
        if (this != &other){
            delete[] _data;                    //记住:这里要删除现有对象的内存数据资源
            _length = other._length;
            _data = new char[_length];         //重新分配内存资源
            std::copy(other._data, other._data + _length, _data);
        }
        return *this;
    }
    //移动构造函数
    MyString(MyString&& other) noexcept: _data(nullptr), _length(0){
        _data = other._data;
        _length = other._length;

        other._data = nullptr;                 //记住:这里要清空源对象数据指针
        other._length = 0;
        //或者
        *this =  std::move(other);
    }
```

```cpp
        //移动赋值运算符
        MyString& operator=(MyString&& other) noexcept{
            if (this != &other){
                delete[] _data;              //记住：这里要删除现有对象的内存数据资源
                _data = other._data;
                _length = other._length;
                other._data = nullptr;       //记住：这里要清空源对象数据指针
                other._length = 0;
            }
            return *this;
        }
        size_t Length() const{
            return _length;
        }
private:
    size_t _length;
    char* _data;
};

int main() {
    vector<MyString> v;
    v.push_back(MyString("hello"));              //调用移动构造函数
    v.push_back(MyString("Modern C++"));         //调用移动构造函数
    v.insert(v.begin() + 1, MyString("!"));      //调用移动构造函数
    return 0;
}
```

3.5.3 引用限定符

在 C++中，函数重载是最常见的程序设计方法。函数重载通常提供左值引用限定重载，但如果未定义右值引用限定重载，则对使用具有右值隐式对象的函数调用都将导致编译错误。为了解决此类问题，C++引入了一个鲜为人知的功能，称为引用限定符。引用限定符提供了一种有用的方法，我们可以根据成员函数是通过左值引用还是右值引用调用的来重载成员函数。通过引用限定符，我们可以创建两个版本的重载函数，一个用于隐式对象是左值引用的情况，另一个用于隐式对象是右值引用的情况。

在下面的示例代码中，我们在类 Student 中定义了成员函数 getStudentName()，该函数返回一个字符串的引用。这就要求使用该函数时采用左值引用，而不能采用右值引用。

【例 3-19】

```cpp
#include <iostream>
#include <string>
#include <string_view>

class Student{
private:
    std::string _name;

public:
```

```cpp
    Student(std::string_view name): _name {name} {}
    const std::string& getStudentName() const { return _name;}
};

//createStudent()通过值返回一个类 Student
Student createStudent(std::string_view name){
    Student s { name };
    return s;
}

int main(){
    //正确：通过右值返回
    std::cout << createStudent("Tom Smith").getStudentName() << '\n';

    //错误：createStudent()的返回对象被销毁后，ref 变成了悬挂引用（dangling reference）
    const std::string& ref {createStudent("Jack Sparks").getStudentName()};
    std::cout << ref << '\n';              //未定义的行为
    return 0;
}
```

通过引用限定符可以解决上面的问题。我们可以将引用限定符（如&和&&）跟在函数getName()声明之后，&表示左值引用，&&表示右值引用。

在 C++中，引用限定符不仅为成员函数调用对象类型（左值引用和右值引用）提供了精细化控制，更能与函数式编程范式深度结合，尤其在涉及引用返回值的函数设计中体现其优势。

下面的示例使用引用限定符对类 Student 进行改造。

【例 3-20】

```cpp
#include <iostream>
#include <string>

class Student{
private:
    std::string _name;

public:
    Student(std::string name): m_name {name} {}
    //&仅匹配隐式左值对象调用
    const std::string& getStudentName() const & {
        return _name;
    }
    //&&仅匹配隐式右值对象调用
    std::string  getStudentName() const && {
        return _name;
    }
};

//createStudent()通过值返回一个类 Student
Student createStudent(const std::string name){
    Student e {name};
    return e;
}
```

```cpp
int main(){
    Student david { "David" };
    //隐式左值对象调用：std::string& getStudentName()const &
    std::cout << david.getStudentName() << '\n';
    //隐式右值对象调用：std::string getStudentName() &&
    std::cout << createStudent("Hansen").getStudentName() << '\n';
    return 0;
}
```

在上面的 main()中，david 是左值对象，通过 david 调用 getStudentName()const &是隐式左值对象调用。而 createStudent("Hansen")创建的是一个隐式右值对象，因此属于隐式右值对象调用 std::string getStudentName() const &&。

3.6 本章小结

本章介绍了 C++中的一些面向对象的技术，主要是 C++11 及以后版本引进的面向对象的技术。这些新技术主要集中在构造函数、析构函数、虚函数、重载等方面。

关键字 explicit 通常放在带有参数的构造函数前，用来阻止通过初始参数转换构造对象。关键字 default 通常放在构造函数的后面，表示默认的构造函数。关键字 delete 通常也跟在构造函数的后面，表示废弃该构造函数，如拷贝构造函数、移动构造函数等。

如果希望程序中使用唯一的全局静态对象，那么需要在类的外部阻止创建类的对象。这时需要将构造函数定义为 private，析构函数也放在 private 部分，同时在拷贝构造函数、移动构造函数后加上 "= delete;"。

纯虚函数是指虚函数名后跟 "=0;"。如果一个类中的所有虚函数后都跟 "=0;"，则该类一般称为抽象类。抽象类是不能被实例化的。纯虚函数在派生类中必须实现。

C++的多态性是通过虚函数体现的。具有虚函数的类都维护了一个虚函数表。虚函数表是仅包含虚函数内存地址的表，并且是按类维护的。在创建具有虚函数的对象时，编译器会向类的对象中添加一个隐藏成员，名为__vptr。该隐藏对象是指向虚函数表的指针，并且按对象实例进行维护，即每个对象都维护一个虚函数表。__vptr 被继承到所有的派生类。如果从基类创建多个派生类，则所有派生类中的__vptr 都指向特定类的同一个虚函数表。

如果虚函数名后跟着 override，则表明该虚函数做了重写；如果虚函数名后跟着 final，则表明虚函数不能被子类重写；如果类名后跟着 final，则表明该类不能被继承。

移动语义与复制语义相对应。移动是指将资源从一个对象移开并搬到另一个对象。复制语义在对象间进行复制，得到复制对象的精确副本，属于左值引用。移动语义通过右值引用实现，不需要在内存中进行拷贝赋值，从而提高了效率。

引用限定符让我们以明确的方式告诉编译器应该调用左值引用函数，还是右值引用函数，从而实现对函数的更精细的控制。

函数调用运算符 operator()用于调用函数和传递参数，它由称为函数对象的类的实例重载。当重载函数调用运算符时，需要创建一个可用于传递参数的运算符函数。在面向对象编程语言中，operator()可以被视为普通的运算符，类的对象可以调用函数 operator()，就像调用赋值运算符一样。

第 4 章
模 板

模板是 C++中的一个简单但功能非常强大的工具。模板最简单的功能是将数据类型作为参数传递，这样就不需要为不同的数据类型编写相同的代码。例如，在需要对不同的数据类型进行排序时，可以编写一个 sort()函数并将数据类型作为参数进行传递，而不是编写和维护多个版本的sort()代码。

模板作为一个 C++实体，主要定义为：
- 函数模板（可以是成员函数）。
- 类模板（可以是嵌套类）。
- 类型别名和别名模板（自 C++11 起）。
- 变量模板（自 C++14 起）。
- 概念和约束（自 C++20 起）。

本章按照上面的顺序安排内容并一一展开介绍。在介绍每种模板时，本章将主要集中在以下的知识点上：
- 函数模板实例化（function template instantiation）。
- 函数模板实参推导（function template argument deduction）。
- 显式函数模板实参（explicit function template argument）。
- 隐式函数模板实参（implicit function template argument）。
- 函数模板实参替换（function template argument substitution）。
- 函数模板实参重载（function template argument overloading）。
- 参数包（parameter pack）。
- 折叠表达式（folding expression）。
- 显式类模板实例化（explicit class template instantiation）。
- 隐式类模板实例化（implicit class template instantiation）。
- 模板形参（template parameter）和模板实参（template argment）。
- 类模板实参推导（class template argument deduction）。
- 非类型模板参数中的类类型（class types in non-type template parameter）。
- 用 auto 声明非类型模板参数（declaring non-type template parameters with auto）。
- 类型别名（type alias）和别名模板（alias template）。
- 变量模板（variable templates）。
- 概念（concept）和约束（constraint）。
- 约束表达式（requires expression）。

4.1 函数模板

函数模板是一系列函数,这些函数有完全相同的逻辑代码,但适应于不同的数据类型,因此可将数据类型抽象为参数(称为模板参数)。函数模板的语法格式如下:

```
template <parameters> function-declaration
template <parameters> requires constraint function-declaration
function-declaration-with-placeholders
```

其中:

(1) parameters:模板的参数列表,非空,用逗号分隔。每个参数要么是非类型模板参数,要么是类型模板参数,要么是模板模板参数(即模板参数嵌套),要么是其中任何一个参数包(从 C++11 开始)。与任何模板一样,参数可能会受到约束(从 C++20 开始)。

(2) function-declaration:声明的函数名称将成为模板名称。

(3) requires constraint:约束表达式,用于限制函数模板可接收的模板参数。

(4) function-declaration-with-placeholders:一个函数声明,其中至少一个参数的类型使用占位符 auto 或 concept auto。

一个简单的函数模板示例如下:

```cpp
template <typename T>                //typename 有时也使用 class
T max(T x, T y){
    return (x < y) ? y : x;
}
```

函数模板本质上是在实例化后编译器生成相应的正确函数。

【例 4-1】

```cpp
#include <iostream>

//声明和定义一个函数模板
template <typename T>
T max(T x, T y){
    return (x < y) ? y : x;
}

int main(){
    //实例化并调用函数模板
    std::cout << max<int>(3, 5) << endl;
    std::cout << max<double>(2.5, 3.5) << endl;
    return 0;
}
```

当编译器遇到 max<int>(3, 5) 和 max<double>(2.5, 3.5) 时会生成下面的两个函数:

```cpp
int max<int>(int x, int y){
    return (x < y) ? y : x;
}
```

```
double max<double>(double x, double y){
    return (x < y) ? y : x;
}
```

上面提到模板参数有：非类型模板参数、类型模板参数、模板模板参数（即模板参数嵌套）或参数包。

类型模板参数的函数模板如下：

```
template<typename T>
T f(T t){ /* ... */ }
```

类型模板参数带有默认值的函数模板如下：

```
template<typename T = double>
T f(T t) { /* ... */ }
```

非类型模板参数的函数模板如下，其中 B 为非类型模板参数，T 为类型模板参数。

```
template<bool B, typename T>
void f(T t){ /* ... */ }
```

模板参数嵌套的情况如下：

```
template<template<class> class P> class X { /* ... */ };
X<A> xa;
```

带有类型模板参数包的函数模板如下：

```
template<class... Targs>
void f(Targs... args) {/* ... */}
```

4.1.1 函数模板实例化

函数模板本身不是类型、函数或任何其他实体，编译器无法从仅仅包含函数模板定义的源文件生成任何代码。为了生成函数，必须实例化函数模板，即必须确定函数模板参数，以便编译器可以从函数模板生成实际函数。函数模板实例化有显式和隐式两种方式：

```
template<typename T>
void f(T s){
    std::cout << s << '\n';
}

template void f<double>(double);    //显式地实例化为 f<double>(double)
template void f<>(char);            //显式地实例化为 f<char>(char)，函数模板参数通过推导获得
template void f(int);               //显式地实例化为 f<int>(int)，函数模板参数通过推导获得
```

函数模板实例化可以采用隐式方式，请看下面的示例代码。

【例 4-2】

```
#include <iostream>

template<typename T>
void f(T s){
```

```
        std::cout << s << '\n';
}

int main(){
    f<double>(1);                   //隐式地实例化并调用 f<double>(double)
    f<>('a');                       //隐式地实例化并调用 f<char>(char)
    f(7);                           //隐式地实例化并调用 f<int>(int)
    void (*pf)(std::string) = f;    //隐式地实例化为 f<string>(string)
    pf("▽");                        //调用 f<string>(string)
    return 0;
}
```

4.1.2 函数模板实参推导

为了实例化函数模板，必须知道函数模板的每个参数，但不必指定每个参数。如果可能，编译器将从实参中推断出缺少的参数，生成正确参数类型的函数。这种情况发生在当编译器尝试调用函数并获取函数模板的地址时。例如：

```
template<typename To, typename From>
To convert(From f);

void g(double d){
    int i = convert<int>(d);        //隐式地实例化为 convert<int,double>(double)
    char c = convert<char>(d);      //隐式地实例化为 convert<char,double>(double)
    int(*ptr)(float) = convert;     //隐式地实例化为 convert<int, float>(float)
    int pi = ptr(3.14159);
}
```

4.1.3 显式函数模板实参

函数模板在实例化和调用时必须指定函数模板实参。函数模板实参可以通过以下方式获取：
- 函数模板实参推导。
- 函数模板参数的默认值。
- 显式指定函数模板实参，这可以在以下的上下文中完成：
    * 在函数调用的表达式中。
    * 当函数的地址被获取时。
    * 对函数的引用被初始化时。
    * 指向成员函数的指针被确定时。
    * 在显式的说明规范中。
    * 在显式的实例化中。
    * 在友元函数声明中。

指定的函数模板实参必须与函数模板形参匹配（即 type 表示类型，non-type 表示非类型，template 表示模板）。实参个数不能多于形参个数（除非一个参数是参数包，在这种情况下，每个非参数包的参数都必须有一个参数）。

如果函数模板形参中含有非类型函数模板参数，则实例化和调用函数模板时，指定的非类型函数模板参数必须与相应的非类型函数模板参数相匹配，或者可以转换为非类型函数模板参数。

不参与函数模板参数推导的函数参数（如显式指定了相应的函数模板参数）会隐式转换为相应函数参数的类型。例如：

```cpp
template<class T>
class Array { /*...*/ };              //类模板

template<class T>                     //函数模板
void sort(Array<T>& v);

template<>                            //实例化为 T = int
void sort(Array<int>&);

template<>
void sort<int>(Array<int>&);          //实例化，但不需要为 sort 指定 int 类型
```

4.1.4 函数模板实参替换

当所有的函数模板实参都已指定、推导或从函数模板参数默认值中获取时，函数参数列表中的函数模板形参在函数调用时都将替换为相应的函数模板实参。我们将此称为函数模板实参替换。如果函数模板实参替换失败（即无法用推导或提供的函数模板实参替换函数模板形参），则会从重载集中删除函数模板。请看下面的例子。

我们声明了三个函数模板：

```cpp
template<class T>
void f(T t);

template<class X>
void g(const X x);

template<class Z>
void h(Z z, Z* zp);
```

下面来调用函数模板：

```cpp
f<int>(1);                //函数类型是 void(int)，t 是 int 类型的
f<const int>(1);          //函数类型是 void(int)，t 是 const int 类型的

g<int>(1);                //函数类型是 void(int)，x 是 const int 类型的
g<const int>(1);          //函数类型是 void(int)，x 是 const int 类型的

h<const int>(1, NULL);    //函数类型是 void(int, const int*)
```

在调用函数模板时，指定的函数模板实参替代了函数模板形参。如果没有显式地指定函数模板实参，则通过推导或从函数模板参数默认值中获取函数模板实参。

4.1.5 函数模板实参重写

首先，函数模板和非函数模板一样，都可能会被重载。非函数模板与实例化的函数模板是不

同的。其次，不同函数模板的实例化始终是彼此不同的，即使具有相同返回类型和相同参数列表的两个函数模板，它们也是不同的，但可以通过其显式模板参数列表进行区分。

我们先看函数模板和非函数模板的重载示例。

【例 4-3】

```cpp
#include <iostream>
using namespace std;

//函数模板
template <typename T>
T min(T x, T y){
    std::cout << "called min<int>(int, int)\n";
    return (x < y) ? x : y;
}

//非函数模板
int min(int x, int y){
    std::cout << "called min(int, int)\n";
    return (x < y) ? x : y;
}

int main(){
    cout << min<int>(3, 5) << '\n';     //调用函数模板 min<int>(int, int)
    cout << min<>(2, 3) << '\n';        //推导出函数模板 min<int>(int, int)
    cout << min(1, 2) << '\n';          //调用非函数模板 min(int, int)
    return 0;
}
```

当使用类型或非类型模板参数的表达式出现在函数参数列表或返回类型中时，该表达式仍是出于重载目的函数模板签名的一部分。例如：

```cpp
template<int I, int J>
A<I+J> f(A<I>, A<J>);        //重载 1

template<int I, int J>
A<I-J> f(A<I>, A<J>);        //重载 2
```

如果两个函数模板的参数在单一定义规则（ODR）下是相同的，则这两个函数模板称为等效函数模板。也就是说，这两个函数模板包含相同的标记序列，其名称通过名称查找解析为相同的实体，但模板参数的名称可能不同。两个等价的函数模板不是函数重载。例如，下面的两个函数模板是等价的。

```cpp
template<int I, int J>
void f(A<I+J>);

template<int K, int L>
void f(A<K+L>);
```

如果出现以下情况，则两个函数模板被视为等价的：
- 它们在同一范围内声明。

- 它们具有相同的函数名称。
- 它们具有等效的模板参数列表，这意味着列表的长度相同，并且对应于每个相应的参数。

4.1.6 参数包

参数包用于接收零个或多个（非类型、类型或模板）的函数模板参数，至少有一个参数包的函数模板称为可变参数函数模板。例如：

```cpp
template<typename... Types>
void g(Types... args);
g();                    //args 不含有实参
g(1);                   //args 含有一个实参，是 int 类型的
g(5, 2.0);              //args 含有两个实参：一个是 int 类型的和一个是 double 类型的

template<typename... Types, typename U, typename=void>
void valid(U u, Types... args);
valid(1.0, 1, 2, 3);    //推导出 U 为 int，Types 为{int, int, int}

template<typename... Types>
void g(Types... args) {}

template<typename... Types>
void f(Types... args){
    g(&args...);        // "&args..." 是一个包扩展，"&args" 是它的模板
}
f(2, 0.2, "hello");
```

【例 4-4】

```cpp
#include <iostream>
template<typename T, typename... Args>
void tprintf(const char* format, T value, Args... args){
    char* p;
    for ( p = format; *p != '\0'; p++) {
        if ( *p == '%' ) {
            std::cout << value;
            tprintf(p+1, args...);          //递归调用
            return;
        }
        std::cout << *p;
    }
}
int main(){
    tprintf("% Modern C++% %\n", "Hello", '!', 123);
    return 0;
}
```

4.1.7 折叠表达式

C++17 引进的折叠表达式具有一项强大的功能，即允许程序员在二元运算符上减少或折叠参数包。引入折叠表达式是为了简化在可变参数函数模板上操作的代码，使代码更加简洁和更具可读性。折叠表达式有以下几种语法：

```
( pack op ... )                              // (1)
( ... op pack )                              // (2)
( pack op ... op init )                      // (3)
( init op ... op pack )                      // (4)
```

其中：op 表示一个二元运算符，如 "+" "-" "*" "/" "%" "^" "&" "|" "=" "<" ">" "<<" ">>" "+=" "-=" "*=" "/=" "%=" "^=" "&=" "|=" "<<=" ">>=" "==" "!=" "<=" ">=" "&&" "||" "," ".*" "->*"；pack 表示一个表达式，包含未展开的参数包，并且不包含优先级低于顶层 const 的运算符；init 表示一个表达式，不包含未展开的参数包和优先级低于顶层 const 的运算符。上述语法中，(1) 和 (2) 是一元折叠表达式的语法，(3) 和 (4) 是二元折叠表达式的语法。下面举例介绍折叠表达式的应用。

```
template <typename... Args>
bool logicalAnd(Args... args) {
    return (true && ... && args);           //二元折叠逻辑"与"
}
bool b = true;
bool& b2 = b;
logicalAnd(b, b2, true);                    //返回值为 true

template<typename... Args>
bool all(Args... args) {
return (... && args);                       //一元折叠逻辑"与"
}
bool b = all(true, true, true, false);      //返回值为 false

template<typename... Args>
bool any(Args... args) {
    return (... || args);                   //一元折叠逻辑"或"
}
bool b = any( false, false, true, false);   //返回值为 true

template <typename... Args>
auto sum(Args... args) {
    return (... + args);                    //一元折叠累加
}
sum(2.0, 3.0f, 5);

template<typename T, typename... Args>
void push_back_vec(std::vector<T>& v, Args&&... args){
    static_assert((std::is_constructible_v<T, Args&&> && ...));
```

```
(v.push_back(std::forward<Args>(args)), ...);
}

std::vector<int> v;
push_back_vec(v, 6, 2, 45, 12);
```

4.2 类模板

类模板定义了一系列的类。和函数模板类似，类模板定义为独立于数据类型的类。标准模板库（STL）中的类都是类模板。类模板的声明语法格式如下：

```
template < parameter-list >
class-declaration
```

这里的参数列表 parameter-list 中的每一个参数可以是非类型模板参数、类型模板参数或模板模板参数。我们先来看看类模板是如何定义和使用的。

【例 4-5】

```
#include <iostream>
using namespace std;

template <typename T>
class MyArray {
private:
    T* pData;
    int size;

public:
    MyArray(T arr[], int n);
    void print();
};

template <typename T>
MyArray<T>::MyArray(T arr[], int n){
    pData = new T[n];
    size = n;
    for (int i = 0; i < size; i++)
    pData[i] = arr[i];
}

template <typename T>
void MyArray<T>::print(){
    for (int i = 0; i < size; i++)
    cout << " " << *(ptr + i);
    cout << end
}

int main(){
```

```cpp
    int arr[5] = { 1, 2, 3, 4, 5 };
    Array<int> a(arr, 5);
    a.print();
    return 0;
}
```

4.2.1 显式类模板实例化

类模板本身不是类型、对象或任何其他实体，编译器不会从仅包含类模板定义的源文件生成任何代码。为了显式生成类的代码，必须实例化类模板，即必须提供类模板实参，以便编译器可以从类模板的定义生成实际的类。

显式类模板实例化的语法格式为：

```
template class-key template-name < argument-list > ;
```

显式类模板实例化强制实例化所引用的类、结构或联合，可以出现在类模板定义之后的代码中。对于给定的参数列表，只允许在整个代码中出现一次显式类模板实例化，并且不需要诊断。显式类模板实例化的示例如下：

```cpp
template<class T>
class Y {
public:
    void mf(){}
};
template class Y <char*>;                //显式类模板实例化
template void Y <double>::mf();          //显式类模板实例化

template<class T>
struct Z {
    void f() {}
void g() {}
};
template struct Z <double>;              //显式类模板实例化
template void Z <double>::f();           //显式类模板实例化
template void Z <double>::g();           //显式类模板实例化
```

4.2.2 隐式类模板实例化

类模板同样可以隐式实例化，特别是在构造类模板的对象时。请看下面的例子。

【例 4-6】

```cpp
template<typename T>
class Apple {
public:
Apple() = default;
    bool IsGreen() const { return false; }
    bool IsRed() const { return true; }
};
```

```cpp
int main(){
    Apple<int> apple;              //隐式实例化 int
    if( apple.IsRed()) {}
    return 0;
}
```

我们再来看一个简单的计算器类模板的示例代码。

【例 4-7】

```cpp
#include <iostream>
using namespace std;

template <class T>
class Calculator{
public:
    Calculator(T n1, T n2) {
        num1 = n1;
        num2 = n2;
    }
    T add() {return num1 + num2;}
    T subtract() {return num1 - num2;}
    T multiply() {return num1 * num2;}
    T divide() {return num1 / num2;}
    void displayResult() {
        cout << "Numbers: " << num1 << " and " << num2 << "." << endl;
        cout << num1 << " + " << num2 << " = " << add() << endl;
        cout << num1 << " - " << num2 << " = " << subtract() << endl;
        cout << num1 << " * " << num2 << " = " << multiply() << endl;
        cout << num1 << " / " << num2 << " = " << divide() << endl;
    }
private:
    T num1, num2;
};

int main(){
    Calculator<int> intCalc(2, 1);              //隐式实例化 int
    Calculator<float> floatCalc(2.4, 1.2);      //隐式实例化 float

    cout << "Int results:" << endl;
    intCalc.displayResult();

    cout << endl << "Float results:" << endl;
    floatCalc.displayResult();
    return 0;
};
```

4.2.3 类模板形参和类模板实参

我们将 template parameter 称为类模板形参,是类模板在声明和定义时的参数;将 template argment 成为类模板实参,是类模板在实例化时使用的参数。

类模板形参列表中的每个参数可以是非类型模板参数、类型模板参数、模板模板参数。模板模板参数既是模板参数也是模板。这里我们只介绍非类型模板参数和类型模板参数。

(1)非类型模板参数主要有以下一些基本数据类型:
- 值引用(value reference)。
- 空指针(nullptr)。
- 指针(pointer)。
- 枚举类型(enumerator)。
- 整型(int)。

非类型模板参数的示例如下:

```
template<int N>
struct S {
    int a[N];
};

template<const char*>
struct S2 {};
```

(2)类型模板参数:对于类型模板参数,只有在实例化时才能确定具体的模板参数类型。类型模板参数的示例如下:

```
template<class T>                    //不带有默认值的类型模板参数
class My_vector { /* ... */ }

template<class T = void>             //带有默认值的类型模板参数
struct My_op_functor { /* ... */ };

template<typename... Ts>             //模板参数包
class My_tuple { /* ... */ };
```

在实例化类模板时,每个类模板形参(非类型模板参数、类型模板参数或模板模板参数)都必须替换为相应的类模板实参。实参要么显式提供,要么有默认值,要么从初始值设定项推导出来(从 C++17 开始)。

```
template<typename T>
class X {};

struct A{};
int main(){
    X<A> x1;            //正确:A 是类型名
    X<A*> x2;           //正确:A*是类型名
}
```

下面介绍类模板实参的类型推导。

4.2.4 类模板实参推导

类模板实参推导与函数模板实参推导的方式非常相似，只是类模板包括类构造函数。例如：

```
template <typename T = float>
struct MyContainer {
    T val;
    MyContainer () : val{} {}
    MyContainer (T val) : val{val} {}
    //...
};

MyContainer c1 {1};         //正确，实例化为 MyContainer<int>
MyContainer c2;             //正确，使用默认的 float 实例化 MyContainer<float>
```

4.2.5 非类型模板参数中的类类型

在 C++20 之前，非类型模板参数的类型限制为基本数据类型，如 bool、int 等。从 C++20 起，类模板的参数列表中可以出现类类型的非类型模板参数，被传入作为类模板实参的对象可以是 const T 类型的，其中 T 是对象的类型，并且具有静态存储持续时间。

```
//先定义一个类类型
struct foo {
    foo() = default;
    constexpr foo(int) {}
};

template <foo f>                    //f 为类类型的非类型模板参数
auto get_foo() {
  return f;
}
get_foo();                          //使用 foo 的默认构造函数
get_foo<foo{123}>();                //使用 foo 的参数构造函数
```

4.2.6 用 auto 声明非类型模板参数

将非类型模板参数声明为 auto 类型，按照 auto 类型的推导规则，可以从类模板实参类型中推断出类模板形参类型。例如：

【例 4-8】

```
#include <iostream>
template <auto N>                   //定义非类型模板参数
void print(){
    std::cout << N << '\n';
}
```

```
int main(){
    print<5>();                //N 被推导为 int 类型
    print<'c'>();              //N 被推导为 char 类型
    return 0;
}

template <auto... seq>
struct my_integer_sequence{
    //...
};
```

显式传入 int 类型作为类模板实参，例如：

```
auto seq1 = std::integer_sequence<int, 0, 1, 2>();      //推导出的类型为 int
auto seq2 = my_integer_sequence<0, 1, 2>();             //推导出的类型为 int
```

4.3 类型别名和别名模板

类型别名是引用先前定义的类型（类似于 typedef）的名称。别名模板是引用一系列类型的名称。类型别名和别名模板具有以下的声明语法：

```
//类型别名
using identifier attr (optional) = type-id;
//别名模板
template < template-parameter-list >
using identifier attr (optional) = type-id;
```

其中：type-id 可能会引入新类型，但不能直接或间接引用标识符；identifier 是类型别名声明引入的一个名称，该名称可用作为 type-id 表示的类型的同义词，但它不是引入了新类型，也不是更改现有类型名称的含义。类型别名声明和 typedef 声明之间没有区别。类型别名声明可能出现在块作用域、类作用域或命名空间作用域中。

别名模板是一个模板，当专用化时，它等效于将别名模板的模板参数替换为 type-id 中的模板参数的结果。下面用代码进一步介绍类型别名和类型别名模板。

```
template<class T>                        //类模板
struct Alloc {};

template<class T>                        //别名模板
using Vec = vector<T, Alloc<T>>;         //type-id 是 vector<T, Alloc<T>>

Vec<int> v;                              //Vec<int>等价于 vector<int, Alloc<int>>
```

下面来看一个综合的例子。

【例 4-9】

```
#include <iostream>
#include <string>
#include <type_traits>
```

```cpp
#include <typeinfo>

//类型别名，等同于 typedef std::ios_base::fmtflags flags;
using flags = std::ios_base::fmtflags;
flags fl = std::ios_base::dec;

//类型别名，等同于 typedef void (*func)(int, int);
using func = void (*) (int, int);

//现在 func 代表指向函数的指针
void example(int, int) {}
func f = example;

//别名模板
template<class T>
using ptr = T*;

//现在 ptr<T>是一个类型别名，是指向类型 T 的指针的别名
ptr<int> x;              //ptr<int>是指向整型的指针类型，x 是该类型的对象。

//类型别名，用来隐藏一个类模板参数
template<class CharT>
using mystring = std::basic_string<CharT, std::char_traits<CharT>>;

//str 是类型 std::basic_string<char, std::char_traits<char>>的对象
mystring<char> str;

//类型别名可以引进一个 typedef 名，用于类成员
template<typename T>
struct Container { using value_type = T; };

//函数模板
template<typename ContainerT>
void info(const ContainerT& c){
    typename ContainerT::value_type T;
    std::cout << "ContainerT is `" << typeid(decltype(c)).name() << "`\n"
                 "value_type is `" << typeid(T).name() << "`\n";
}

//类型别名用于简化 std::enable_if 的语法
template<typename T>
using Invoke = typename T::type;

template<typename Condition>
using EnableIf = Invoke<std::enable_if<Condition::value>>;

template<typename T, typename = EnableIf<std::is_polymorphic<T>>>
int fpoly_only(T) { return 1; }
```

```cpp
struct S { virtual ~S() {} };

int main(){
    Container<int> c;
    info(c);                //Container::value_type 在函数中是整型
    fpoly_only(c);          //错误：enable_if 禁止这样做
    S s;
    fpoly_only(s);          //正确：enable_if 允许这样做
    return 0;
}
```

4.4 变量模板

C++14 引入了变量模板。变量模板允许创建一组可自定义的变量或静态数据成员，从而为代码提供灵活性。变量模板的语法格式如下：

```
template < parameter-list >
variable-declaration
```

从变量模板实例化的变量称为实例化变量，从静态数据成员模板实例化的静态数据成员称为实例化的静态数据成员。例如：

```cpp
template<class T>                       //变量模板
constexpr T pi = T(3.1415926535897932385L);

template<class T>
T circular_area(T r)                    //函数模板
{
    return pi<T> * r * r;               //pi<T>是变量模板实例化
}
```

与其他静态成员一样，我们可能需要定义静态数据成员模板。此类定义在类定义之外提供。在命名空间范围内，静态数据成员的模板声明也可以是类模板的非模板数据成员的定义。

```cpp
struct limits{
    template<typename T>
    static const T min;                 //声明一个静态数据成员模板
};

template<typename T>
const T limits::min = { };              //定义一个静态数据成员模板，即初始化

template<class T>
class X{
    static T s;                         //声明一个非模板的静态成员数据，因为T是类模板参数
};

template<class T>
```

```cpp
T X<T>::s = 0;                      //非模板的静态成员数据初始化，注意 T 为类模板参数

using namespace std::literals;
struct matrix_constants{
    template<class T>
    using pauli = hermitian_matrix<T, 2>;        //别名模板

    template<class T>                            //静态数据成员模板
    static constexpr pauli<T> sigmaX = {{0, 1}, {1, 0}};

    template<class T>                            //静态数据成员模板
    static constexpr pauli<T> sigmaY = {{0, -1i}, {1i, 0}};

    template<class T>                            //静态数据成员模板
    static constexpr pauli<T> sigmaZ = {{1, 0}, {0, -1}};
};
```

【例 4-10】

```cpp
#include <iostream>

template <typename T>
constexpr T Pi{3.14159265358979323846264338327};

struct NewType {
    constexpr NewType(double init)
        : value{int(init)} {}
    int value;
};

int main() {
    auto container{Pi<NewType>};
    std::cout << "container value: " << container.value;
    return 0;
}
```

4.5 概念与约束

C++20 引进了概念与约束。类模板、函数模板和类模板的非模板成员函数都可能与约束相关联。约束指定了对模板实参的要求，实参可用于选择最合适的函数重载和模板专用化。概念是一组要求的命名，要求集的名字称为概念。每个概念都是一个谓词，在编译时计算，成为模板接口的一部分，在模板中用作约束。概念的定义必须出现在命名空间范围内。

4.5.1 概念

概念的定义具有以下形式：

```
template <template-parameter-list>
concept concept-name attr(optional) = constraint-expression;
```

其中:constraint-expression 可以是一个约束或者约束组合,也可以是一个约束表达式。下面的代码定义了一个概念 Derived:

```
template<class T, class U>
concept Derived = std::is_base_of<U, T>::value;          //约束子句
```

下面的代码定义了一个概念 Hashable,任何类型 T 都满足该概念,因此,对于类型为 T 的值 a,约束表达式 std::hash<T>{}(a)被编译转换为 std::size_t。

```
template<typename T>
concept Hashable = requires(T a){                         //约束表达式
    {std::hash<T>{}(a)}-> std::convertible_to<std::size_t>;
};
```

现在应用概念 Hashable 限制模板函数,代码如下:

```
template<typename T>
requires Hashable<T>                                      //约束子句
void f(T) {}
```

也可以这样写:

```
template<typename T>
void f(T) requires Hashable<T> {}                         //函数名后跟着约束子句
```

或者这样写:

```
void f(Hashable auto /*parameterName*/) {}
```

或者这样写:

```
template<Hashable T>
void f(T) {}
```

概念不能递归引用自己,也不能被限制,例如:

```
template<typename T>
concept V = V<T*>;                                        //错误:递归概念

template<class T>
concept C1 = true;

template<C1 T>
concept Error1 = true;                                    //错误:C1 T 试图限制概念

template<class T> requires C1<T>
concept Error2 = true;                                    //错误:约束子句试图限制概念
```

【例 4-11】

```
#include <iostream>
#include <type_traits>
```

```cpp
using namespace std;

//函数模板使用约束子句作为约束
template <typename T>
requires is_integral_v<T>
void print_integer(T value)
{
    cout << "The integer value is: " << value << endl;
}

//定义概念
template <typename T>
concept Printable = requires(T value)
{
    cout << value << endl;
};

//函数模板使用概念 Printable 作为约束
template <Printable T>
void print(T value)
{
    cout << "The printable value is: " << value << endl;
}

int main() {
    print_integer(42);
    print("Hello, World!");
    return 0;
}
```

4.5.2 约束

约束是一组按顺序的逻辑操作和操作数的序列，用于指定对模板参数的要求。约束可以出现在约束表达式中，也可以直接作为概念主体出现。约束通常有三种：与（conjunction）、或（disjunction）、原子约束（atomic constraint）。

（1）与约束：两个约束的与是通过在约束表达式中使用运算符&&形成的。例如：

```cpp
template<class T>
concept Integral = std::is_integral<T>::value;
template<class T>
concept SignedIntegral = Integral<T> && std::is_signed<T>::value;
template<class T>
concept UnsignedIntegral = Integral<T> && !SignedIntegral<T>;
```

仅当满足两个约束时，才满足两个约束的与。与约束从左到右计算并短路，如果不满足左约束，则不尝试将模板参数替换为右约束。例如：

```cpp
template<typename T>
```

```
constexpr bool get_value() { return T::value; }

template<typename T>
requires (sizeof(T) > 1 && get_value<T>())     //两个约束在约束表达式中
void f(T);                                      //1

void f(int);                                    //2

void g(){
    f('A');
}
```

在上述代码中，g()调用 f(int)，当检查约束 1 时，发现 sizeof(char)>1 不满足，因此不再检查 get_value<T>()。

（2）或约束：两个约束的或是通过在约束表达式中使用运算符||形成的。如果满足任一约束，则满足两个约束的或。或约束从左到右计算析取并短路，如果满足左约束，则不尝试将模板参数替换为右约束。例如：

```
template<class T = void>
requires EqualityComparable<T> || Same<T, void>
struct equal_to;
```

（3）原子约束：原子约束由约束表达式 E 和 E 中的模板形参到模板实参的映射形成，也称为参数映射。通过将参数映射和模板实参替换到 E 中可检查原子约束的满足程度。如果替换导致无效的类型或约束表达式，则约束不满足。否则，在任何左值到右值的转换之后，E 应是 bool 类型的 prvalue 常量表达式，并且当且仅当它的计算结果为 true 时，约束才满足。例如：

```
template<typename T>
struct S{
    constexpr operator bool() const { return true; }
};

template<typename T>
requires (S<T>{})
void f(T);                      //1

void f(int);                    //2
void g(){
    f(0);                       //错误：当检查 1 时，S<int>{}不是 bool 类型的
}
```

如果两个原子约束在代码级是由相同的表达式构成的，并且它们的参数映射是等效的，则称两个原子约束是相同的。例如：

```
template<class T>
constexpr bool is_meowable = true;

template<class T>
constexpr bool is_cat = true;
```

```cpp
template<class T>
concept Meowable = is_meowable<T>;

template<class T>
concept BadMeowableCat = is_meowable<T> && is_cat<T>;

template<class T>
concept GoodMeowableCat = Meowable<T> && is_cat<T>;

template<Meowable T>
void f1(T);                         //1

template<BadMeowableCat T>
void f1(T);                         //2
template<Meowable T>void
f2(T);                              //3

template<GoodMeowableCat T>
void f2(T);                         //4

void g(){
    f1(0);
    f2(0);
}
```

在上面的代码中，f1(0) 调用是错误的，因为存在歧义，is_meowable<T>在 Meowable 和 BadMeowableCat 构成了不同的原子约束；f2(0)调用是正确的。

4.5.3 约束表达式

约束表达式是 C++20 引进的，它会生成一个 bool 类型的 prvalue 表达式，该表达式描述约束。约束通过概念的形式来限制作用域中的模板参数、参数列表中的参数，以及从封闭上下文中可见的任何其他声明。约束表达式的语法格式如下：

```
requires { requirement-seq }
requires ( parameter-list(optional)) {requirement-seq}
```

下面我们举例介绍约束表达式的应用。

```cpp
template<typename T>
concept Addable = requires (T a, T b){     //约束表达式
    a+b;
};

template<typename T>
requires Addable<T>                        //这里是约束子句，不是约束表达式
auto add(T a, T b){
    return a + b;
}
```

```cpp
template<typename T>                    //函数名称后跟着约束子句
auto add(T a, T b) requires Addable<T> {
    return a + b;
}

template<Addable T>                     //限定模板参数
auto add(T a, T b){
    return a + b;
}

//简约函数模板
auto add(Addable auto a, Addable auto b) {
    return a + b;
}
```

从上面的应用可以看出，约束表达式一般先用于定义概念，然后通过概念以约束子句的形式出现在函数模板的定义中，用于对模板参数进行限制。

在约束表达式中，requirement-seq 是一组顺序的要求，每个要求可以是简单要求（simple requirement）、类型要求（type requirement）、复合要求（compound requirement）或嵌套要求（nested requirement）。

（1）简单要求的示例如下：

```cpp
template<typename T>
concept Addable = requires (T a, T b){
    a + b;                  //简单要求
};

template<class T, class U = T>
concept Swappable = requires(T&& t, U&& u){
    swap(std::forward<T>(t), std::forward<U>(u));
    swap(std::forward<U>(u), std::forward<T>(t));
};
```

（2）类型要求的示例如下：

```cpp
template<typename T>
using Ref = T&;

template<typename T>
concept C = requires{
    typename T::inner;      //要求嵌套的成员名
    typename S<T>;          //要求类模板明细
    typename Ref<T>;        //要求别名模板替换
};

template<class T, class U>
using CommonType = std::common_type_t<T, U>;

template<class T, class U>
```

```cpp
concept Common = requires (T&& t, U&& u){
    typename CommonType<T, U>;
    { CommonType<T, U>{std::forward<T>(t)} };
    { CommonType<T, U>{std::forward<U>(u)} };
};
```

（3）复合要求的示例如下：

```cpp
template<typename T>
concept C2 = requires(T x){
    {*x} -> std::convertible_to<typename T::inner>;
    {x + 1} -> std::same_as<int>;
    {x * 1} -> std::convertible_to<T>;
};
```

（4）嵌套要求的示例如下：

```cpp
template<class T>
concept Semiregular = requires(T a, std::size_t n){
    requires Same<T*, decltype(&a)>;{ a.~T()} noexcept;
    requires Same<T*, decltype(new T)>;
    requires Same<T*, decltype(new T[n])>;      //嵌套要求
    { delete new T };                            //复合要求
    { delete new T[n]};                          //复合要求
};
```

下面的代码将几种要求放在一起组成一个综合的要求。

```cpp
template <typename T>
concept Fooable = requires(T a){
    //简单要求
    a++;                                         //要求能够后++
    ++a;                                         //要求能够前++

    //类型要求
    typename T::value_type;                      //有内部类型成员值类型

    //复合要求
    {a+1}->std::convertible_to<T>;               //a+1 是有效表达式并且可以转成类型 T

    //嵌套要求
    requires std::same_as<T*, decltype(&a)>;
};
```

【例 4-12】

```cpp
struct point{
    int x;
    int y;
};

std::ostream& operator<<(std::ostream& os, point const& p){
```

```cpp
        os << '(' << p.x << ',' << p.y << ')';
        return os;
}

template <typename T>
constexpr bool always_false = std::false_type::value;

template <typename T>
std::string as_string(T a){
    constexpr bool has_to_string = requires(T x){
        { std::to_string(x) } -> std::convertible_to<std::string>;
    };

    constexpr bool has_stream = requires(T x, std::ostream& os){
        {os << x} -> std::same_as<std::ostream&>;
    };

    if constexpr (has_to_string){
        return std::to_string(a);
    }
    else if constexpr (has_stream){
        std::stringstream s;
        s << a;
        return s.str();
    }
    else
    static_assert(always_false<T>, "The type cannot be serialized");
}

int main(){
    std::cout << as_string(42) << '\n';
    std::cout << as_string(point{1, 2}) << '\n';

    //错误：类型不能被序列化
    std::cout << as_string(std::pair<int, int>{1, 2}) << '\n';
    return 0;
}
```

4.6 本章小结

本章较为详细地介绍了 C++ 中的模板，并重点介绍了自 C++11 起引进的一些有关模板的重要特性。模板是 C++ 的强大功能之一，也是 C++ 初学者感到困难的地方。在软件开发中，熟练使用模板不仅可以大大减少代码量，提高开发效率和软件的可维护性，还有助于培养自己的编程风格，提高自己的编程水平。

在函数模板部分，本章主要介绍了函数模板实例化、函数模板实参推导、显式函数模板实参、函数模板实参替代、函数模板实参重写、参数包、折叠表达式等内容。通过学习这些内容，读者可以全面掌握函数模板从声明定义到实例化和调用的所有技术。

类模板与函数模板一样，在声明后需要进行实例化。本章重点介绍了显式类模板实例化、隐式类模板实例化、模板形参和模板实参、非类型模板参数中的类类型、用 auto 声明非类型模板参数。

类型别名和别名模板也是本章的一个知识点。类型别名是引用先前定义的类型（类似于 typedef）的名称，使用关键字 using 可命名类型别名。别名模板是引用一系列类型的名称，在定义时需要使用关键字 template 和 using。

变量模板是 C++14 引入的，允许创建一组可自定义的变量或静态数据成员，从而为代码提供灵活性。变量模板定义了一系列变量或静态数据成员，因此在声明定义变量模板后需要实例化才能使用。从变量模板实例化的变量称为实例化变量，从静态数据成员模板实例化的静态数据成员称为实例化的静态数据成员。

约束与概念是 C++20 引进的。约束主要应用于类模板、函数模板和类模板的非模板成员函数，指定了对模板参数的要求。在概念定义中使用约束，并通过概念和约束子句的形式，可将约束用于限制模板参数，从而选择最合适的函数重载和模板专用化。

模板在声明时所使用的参数称为模板形参，模板在实例化时传递给模板的参数称为模板实参。不论函数模板，还是类模板，模板本身不是任何实体，编译器不能从仅包含模板定义的源文件生成任何代码，模板必须在实例化后才能生成具体的函数或类。

第 5 章
STL 容器

容器是 STL 的重要组成部分，是存储其他对象的对象。容器被设计并实现为类模板，模板参数为被存储对象（即元素）的类型。容器管理其元素的存储空间，并提供成员函数来直接访问或者通过迭代器来访问元素。C++的容器库是类模板和算法的通用集合，允许程序员轻松实现常见的数据结构，如队列、列表和堆栈等。STL 中的容器通常可划分为以下几类：
- 顺序容器：包括 array、vector、list、forward_list 和 deque。
- 关联容器：包括 set、multiset、map 和 multimap。
- 无序关联容器：包括 unordered_set、unordered_map、unordered_multisetset 和 unordered_multisetmap。
- 容器适配器：包括 stack、queue 和 priority_queue。

根据本贾尼·斯特劳斯特卢普（Bjarne Stroustrup）的说法，容器的设计需要满足两个标准：在单个容器的设计中提供最大的自由度，同时允许容器向用户提供通用接口。这允许在容器的实现中实现最佳效率，并使用户能够编写独立于所使用的特定容器的代码。

5.1 顺序容器

在顺序容器中，array、vector 和 deque 采用的是基于索引的数组结构，在编译时会清楚地设置容器的大小。这意味着 array、vector 和 deque 在本质上是静态的，创建速度很快，但如果有大量的空间未使用，则可能会浪费这些空间。尽管 array、vector 和 deque 的大小最初是固定的，但可以调整大小，以便在数组超过其初始大小时插入元素。由于数组大小是在编译过程中设置的，因此对数组进行动态扩展是一种性能权衡。检索数组的元素很容易，可以通过索引号快速引用任何元素，因此，当获取最后一个元素时不需要遍历所有的中间元素。list 和 forward_list 采用的是一种链表形式的数据结构，其中的节点包含数据元素和指向下一个节点的引用，而上一个元素和下一个元素都实现为双向链表。链表仅在需要时动态创建节点，因此不会浪费未用的空间。这显然是对空间的有效管理，但在存储每个新元素前必须创建一个新节点，会消耗额外的时间；另外，检索过程也相对较慢，因为与采用索引结构的数组不同，链表无法在元素之间跳转。节点是动态创建的，并分配仅由节点的引用元素链接的非顺序内存位置，因此要选择一个元素，必须遍历它之前的所有节点。在最坏的情况下，可能需要遍历除元素所在节点之外的所有节点。

顺序容器的共性成员函数如表 5.1-1 所示。

表 5.1-1　顺序容器的共性成员函数（X 表示容器类型，a 表示容器对象）

表 达 式	类 型	效 果
X(il)	X	等价于 X(il.begin(), il.end())
a = il	X&	使用赋值语句，把 il 拷贝构造赋值给 a
a.emplace(p, args)	iterator	在 p 前插入一个 T 类型的对象
a.insert(p, t)	iterator	在 p 前插入 t 的拷贝
a.insert(p, rv)	iterator	在 p 前插入一个 rv 的拷贝，可能使用移动语义
a.insert(p, n, t)	iterator	在 p 前插入 n 个 t 的拷贝
a.insert(p, i, j)	iterator	在 p 前插入元素[i, j)
a.insert(p, il)	iterator	等价于 insert(p, il.begin(), il.end())
a.erase(q)	iterator	删除元素指针 q
a.erase(q1, q2)	iterator	删除[q1, q2)内的元素
a.clear()	void	销毁 a 中的所有元素
a.assign(i, j)	void	用[i, j)内的元素拷贝进行替换
a.assign(il)	void	等价于用 il.begin()和 il.end()进行赋值

顺序容器中的可选项函数如表 5.1-2 所示。

表 5.1-2　顺序容器中的可选项函数（a：容器对象）

表 达 式	返回类型	效 果	容 器
a.front()	引用	返回*a.begin().	所有顺序容器
a.back()	引用	等价于 auto tmp = a.end()	array、deque、list、vector
a.emplace_front(args)	void	构造一个 T，并放在数组前端	deque、forward_list、list
a.emplace_back(args)	void	构造一个 T，并放在数组尾部	deque、list、vector
a.push_front(t)	void	前端放置 t 的拷贝或移动拷贝	deque、forward_list、list
a.push_back(t)	void	在尾部放置 t 的拷贝或移动拷贝.	deque、list、vector
a.pop_front()	void	去掉第一个元素	deque、forward_list、list
a.pop_back()	void	去掉最后一个元素	deque、list、vector
a[n]	引用	等价于返回*(a.begin() + n);.	array、deque、vector
a.at(n)	引用	返回*(a.begin() + n)	array、deque、vector

5.1.1　std::array

std::array 是一个封装固定大小数组的容器，其声明如下：

template<class T, std::size_t N>
struct array;

容器 std::array 是一种聚合类型，其底层实现等效于一个仅包含 C 语言数组 T[N]作为唯一非静态数据成员的结构体。与 C 语言的数组不同的是，容器 std::array 不会自动隐式转换为指向首元素的指针 T*。作为一种聚合类型，容器 std::array 可以通过最多 N 个可转换为 T 类型的初始值设定项进行初始化，如 std::array<int,3> a = {1,2,3}。

容器 std::array 的优点是将 C 语言数组的性能、可访问性与标准容器的特性相结合，如知道自己的大小、支持赋值、随机访问迭代器等。例如：

```
std::array<int, 5> a1{{3, 4, 5, 1, 2}};
std::array<int, 5> a2 = {1, 2, 3, 4, 5};
std::array<string, 2> a3 = {{string("a"), "b"}};
```

注意：使用容器 std::array 时需要加上#include <array>。

【例 5-1】

```
#include <iostream>
#include <array>
#include <algorithm>

int main (){
    std::array<int,5> a = { 2, 16, 77, 34, 50 };
    std::cout << "a contains:";
    for ( auto it = a.begin(); it != a.end(); ++it )
        std::cout << ' ' << *it;                    //使用迭代器访问数组
    std::cout << '\n';
    std::sort(a.begin(), a.end());
    std::cout << "after sort a contains:";
    for (size_t i = 0; i < a.size(); ++i )
        std::cout << ' ' << a.at(i);                //或者 a[i]
    std::cout << '\n';
    std::cout << "the min element: " << a.front() << '\n';
    std::cout << "the max element: " << a.back() << '\n';
    return 0;
}
```

5.1.2　std::vector

C++中的 std::vector 是一个包含向量容器及其成员函数的类模板，它是在<vector>头文件中定义的。std::vector 的类成员函数为向量容器提供各种功能。std::vector 的声明如下：

```
template< class T, class Allocator = std::allocator <T>>
class vector;

namespace pmr {
    template< class T > using vector = std::vector<T, std::pmr::polymorphic_allocator <T>>;
}
```

std::vector 与动态数组相同，能够在插入或删除元素时自动调整自身大小，其存储由容器自动处理。std::vector 中元素被放置在连续的存储空间中，以便可以使用迭代器访问和遍历元素。在 std::vector 中插入或删除元素的时间复杂度与操作的位置密切相关。在末尾插入元素时需要不同的时间，因为可能需要扩展数组；删除最后一个元素需要恒定的时间，因为不需要调整数组大小。在开头和中间插入或删除元素的时间复杂度是线性的。

元素是连续存储的，这意味着不仅可以通过迭代器访问元素，还可以使用指向元素的常规指针的偏移量来访问元素。指向 std::vector 元素的指针可以传递给任何需要指向数组元素的指针函数。例如：

【例 5-2】

```cpp
#include <vector>
#include <iostream>
using namespace std;

int main(){
    vector<int> v;
    v.assign(5, 10);                //指定 vector 的大小为 5，并为每个元素赋值 10
    for (int i = 0; i < v.size(); i++)
        cout << v[i] << " ";        //访问每个元素

    v.push_back(15);                //在末尾添加 15
    v.pop_back();                   //删除末尾的元素
    v.insert(v.begin(), 5);         //在开头插入新元素
    v.erase(v.begin());             //删除头部的元素
    v.emplace(v.begin(), 5);        //在开头插入新元素
    v.emplace_back(20);             //在末尾插入新元素
    n = v.size();
    for (int i = 0; i < n; i++)
        cout << v[i] << " ";        //访问每个元素

    v.clear();                      //清空容器
    vector<int> v1, v2;
    v1.push_back(1);
    v1.push_back(2);
    v2.push_back(3);
    v2.push_back(4);
    v1.swap(v2);                    //交换两个容器
    return 0;
}
```

需要注意的是，当 std::vector 中的元素是指针对象时，需要手动删除 std::vector 中的指针对象，否则会造成内存泄漏。例如：

【例 5-3】

```cpp
#include <vector>

class A {
public:
    A(int a);
    ~A();
private:
    int a_;
};

int main() {
```

```cpp
std::vector<A*>vec;
vec.push_back(new A(2));
vec.push_back(new A(4));
vec.push_back(new A(5));
vec.push_back(new A(7));

for (std::vector<A*>::iterator it=vec.begin(); it != vec.end(); ++it)
{
    A* a = *it;
    delete a;
    a = NULL;
}
vec.clear();
return 0;
}
```

使用智能指针时,无须手动删除 std::vector 中的对象。上面的代码可变成:

【例 5-4】

```cpp
#include <vector>
#include <utility>

class A {
public:
    A(int a);
    ~A();
private:
    int a_;
};

typedef std::unique_ptr<A> APtr;
std::vector< APtr > vec;

int main() {
    vec.push_back(std::move(new A(2)));
    vec.push_back(std::move(new A(4)));
    vec.push_back(std::move(new A(5)));
    vec.push_back(std::move(new A(7)));
    vec.clear();
    return 0;
}
```

5.1.3　std::list

std::list 是一个容器,支持在其中的任何位置持续插入和删除元素,但不支持快速随机访问。std::list 通常以双向链表的形式实现。与 std::forward_list 相比,std::list 提供了双向迭功能。std::list 的声明如下:

```cpp
template<class T,class Allocator = std::allocator<T>>
```

```cpp
class list;

namespace pmr {
    template< class T >
    using list = std::list<T, std::pmr::polymorphic_allocator<T>>;
}
```

std::list 是允许分配非连续内存的顺序容器。与 std::vector 相比，std::list 的遍历速度较慢，但一旦找到位置，插入和删除元素的速度就会很快且时间恒定。通常，当我们提到 std::list 时，谈论的是一个双向链表。为了实现单向链表，我们使用 std::forward_list。在 std::list 中添加、删除和移动元素不会使迭代器或引用失效。仅当删除相应的元素时，迭代器才会失效。例如：

【例 5-5】

```cpp
#include <iostream>
#include <iterator>
#include <list>

void printlist(list<int> g){
    for (std::list<int>::iterator it = g.begin(); it != g.end(); ++it)
        std::cout << *it << ' ';
}

int main(){
    std::list<int> l1, l2;
    for (int i = 0; i < 10; ++i) {
        l1.push_back(i * 2);
        l2.push_front(i * 3);
    }
    printlist(l1);
    std::cout << endl;
    printlist(l2);

    std::cout << "\nl1.front() : " << l1.front();
    std::cout << "\nl1.back() : " << l1.back();

    std::cout << "\nl1.pop_front() : ";
    l1.pop_front();
    printlist(l1);

    std::cout << "\nl2.pop_back() : ";
    l2.pop_back();
    printlist(l2);

    std::cout << "\nl1.reverse() : ";
    l1.reverse();
    printlist(l1);

    std::cout << "\nl2.sort(): ";
    l2.sort();
```

```
    printlist(l2);

    return 0;
}
```

同样，当 std::list 中的元素是指针对象时，需要手动删除 std::list 中的指针对象，否则会造成内存泄漏。使用智能指针时，无须手动删除 std::list 中的指针对象。请看下面的代码：

【例 5-6】

```cpp
#include <iostream>
#include <list>
struct mystruct {
    int a {};
    mystruct() {}
    mystruct(int aa) : a(aa) { std::cout << "construct " << a << '\n'; }

    mystruct(const mystruct&) = default;
    mystruct& operator=(const mystruct&) = default;

    ~mystruct() {
        std::cout << "destruct " << a << '\n';
    }
};

int main(){
    std::list<mystruct*> mylist;

    mylist.push_back(new mystruct(1));
    mylist.push_back(new mystruct(2));
    mylist.push_back(new mystruct(3));
    mylist.push_back(new mystruct(4));

    mylist.erase(mylist.begin());          //有问题，会造成内存泄漏

    for (auto& a : mylist)
        delete a;                          //需要手动删除 std::list 中的指针对象或者使用智能指针
    return 0;
}
```

5.1.4 std::forward_list

std::forward_list 是一个容器，支持在其中的任何位置快速插入和删除元素，不支持快速随机访问元素。std::forward_list 是单向链表。与 std::list 相比，std::forward_list 在不需要双向迭代时更加节省存储空间。std::forward_list 的声明如下：

```cpp
template<class T,class Allocator = std::allocator<T>>
class forward_list;

namespace pmr{
```

```cpp
    template< class T >
    usingforward_list = std::forward_list<T,std::pmr::polymorphic_allocator<T>>;
}
```

下面的示例展示了 std::forward_list 的创建和初始化。

【例 5-7】

```cpp
#include <iostream>
#include <string>
#include <forward_list>

template<typename T>
std::ostream&operator<<(std::ostream&s,const std::forward_list<T>& v){
    s.put('{');
    for (char comma[]{'\0', ' ', '\0'}; const auto& e : v)
        s << comma << e, comma[0] = ',';
    return s << "}\n";
}

int main(){
    std::forward_list<std::string> w1{"C++", "Java", "Python", "C#", "JavaScript"};
    std::cout << "1: " << w1;

    std::forward_list<std::string> w2(w1);
    std::cout << "2: " << w2;

    std::forward_list<std::string>w3(w1.begin(),w1.end());
    std::cout << "3: " << w3;

    std::forward_list<std::string> w4(5, "Go");
    std::cout << "4: " << w4;

    auto const rg = {"Oracle", "DB2", "MySql"};
    #ifdef __cpp_lib_containers_ranges
    std::forward_list<std::string> w5(std::from_range,rg);
    #else
    std::forward_list<std::string> w5(rg.begin(), rg.end());
    #endif
    std::cout << "5: " << w5;
}
```

下面的示例展示了 std::forward_list 的应用，其中的元素为 std::pair<int, double>。

【例 5-8】

```cpp
#include <iostream>
#include <string>
#include <forward_list>
using namespace std;

void print(forward_list<pair<int, double>> &forwardListOfPairs){
```

```cpp
        for (auto currentPair : forwardListOfPairs){
            pair<int, double> p = currentPair;
            cout << "[ ";
            cout << p.get(0) << ' ' << P.get(1);
            cout << ']';
            cout << '\n';
        }
    }

    int main(){
        forward_list<pair<int, double> >forwardListOfPairs;

        pair<int, double> p1;
        tuple1 = make_pair(1, 2.0);
        forwardListOfPairs.push_front(p1);

        pair<int, double> p2;
        p2 = make_pair(2, 3.0);
        forwardListOfPairs.push_front(p2);

        pair<int, double> p3;
        p3 = make_pair(3, 3.0);
        forwardListOfPairs.push_front(p3);

        pair<int, double> p4;
        tuple4 = make_pair(4, 4.0);
        forwardListOfPairs.push_front(p4);

        print(forwardListOfPairs);
        return 0;
    }
```

同样，当 std::forward_list 中的元素是指针对象时，需要手动删除 std::forward_list 中的指针对象，否则会造成内存泄漏。使用智能指针时，无须手动删除 std::forward_list 中的指针对象。

5.1.5 std::deque

std::deque（双端队列）是一种顺序容器，允许在其开头和末尾快速插入和删除元素。此外，在 std::deque 的两端插入和删除元素永远不会造成其余元素的指针或引用无效。std::deque 的声明如下：

```cpp
template<class T, class Allocator = std::allocator<T>>
class deque;

namespace pmr {
    template< class T >
    using deque=std::deque<T, std::pmr::polymorphic_allocator<T>>;
}
```

队列一般只允许在末尾插入元素，在开头删除元素。这就像现实生活中的队列，人们从前面离开，在后面加入。双端队列是队列的一种特例，其两端都可以执行插入和删除元素的操作。与 std::vector 不同，std::deque 中的元素不是连续存储的。std::deque 使用一系列单独分配的固定大小的数组，并带有额外的标记，这意味着对 std::deque 的索引访问必须执行两次指针取消引用，而对 std::vector 的索引访问只执行一次指针取消引用。

std::deque 的存储会根据需要自动扩展和收缩。std::deque 的扩展比 std::vector 的扩展更方便，因为 std::deque 不涉及将现有元素复制到新的位置。另外，std::deque 通常具有较大的最小内存成本，一个即使只包含一个元素的 std::deque 也必须分配完整的内部数组。

【例 5-9】

```cpp
#include <deque>
#include <iostream>
int main(){
    std::deque<int> d = {7, 5, 16, 8};
    d.push_front(13);
    d.push_back(25);
    d.push_back(30);
    d.push_front(15);

    cout << "\nd.size() : " << d.size();
    cout << "\nd.max_size() : " << d.max_size();

    cout << "\nd.at(2) : " << d.at(2);
    cout << "\nd.front() : " << d.front();
    cout << "\nd.back() : " << d.back();

    cout << "\nd.pop_front() : " << d.pop_front();
    cout << "\nd.pop_back() : " << d.pop_back();

    for (int n : d)
        std::cout << n << ' ';
    std::cout << '\n';
}
```

同样，当 std::deque 中的元素是指针对象时，需要手动删除 std::deque 中的指针对象，否则会造成内存泄漏。使用智能指针时，无须手动删除 std::deque 中的指针对象。

5.2 关联容器

关联容器是一个有序容器，它可实现基于 key 的对象快速查找。如果关联容器中的一个 key 最多包含一个元素，则它支持唯一 key。否则，它支持等效 key。在标准模板库（STL）中，关联容器指的是用于实现关联容器的类模板组，这些类模板组用于存储元素，但对其中的元素施加了一些约束。

关联容器的主要成员函数如表 5.2-1 所示。

表 5.2-1　关联容器的主要成员函数（X 为关联容器类型，a 为关联容器的对象）

表 达 式	效 果
X(c)	构造一个空容器，使用 c 作为比较对象
X u	构造一个空容器
X(i, j, c)	构造一个空容器并插入 range[i,j]，c 作为比较对象
X(i, j)	构造一个空容器并插入 range [i, j)
X(il, c)	X(il.begin(), il.end(), c)
X(il)	X(il.begin(), il.end())
a = il	赋值 range [il.begin(), il.end()) 给 a
a.emplace_hint(p,args)	等价于 a.emplace(std::forward<Args>(args)...)
a.insert(p, t)	在尽可能靠近 p 之前的位置插入 t
a.insert(i, j)	将范围[i, j]中的每个元素插入容器
a.insert(il)	a.insert(il.begin(), il.end())
a.extract(k)	删除容器中的值等效于 k 的第一个元素
a.extract(q)	删除迭代器 q 的元素
a.merge(a2)	提取 a2 中的每个元素，并使用 a 的比较对象将其插入 a 中
a.erase(k)	删除容器中值等于 k 的所有元素
a.erase(q)	删除迭代器 q 的元素
a.erase(q1, q2)	删除 range[q1, q2)内的所有元素
a.clear()	删除容器中的所有元素
a.find(k)	查找容器中值为 k 的元素
a.count(k)	统计容器中值为 k 的元素个数
a.contains(k)	返回 a_tran.find(k) != a_tran.end();
a.lower_bound(k)	迭代器指向第一个元素，其值不小于 k
a.upper_bound(k)	迭代器指向第一个元素，其值不大于 k
a.equal_range(k)	返回由 lower_bound(k)和 upper_bound(k)构成的 pair

5.2.1　std::set

std::set 是一种关联容器，其中的每个元素都必须是唯一的，因为元素的值可以标识元素。元素的值按特定的顺序存储，即升序或降序。在 std::set 中搜索、删除和插入元素的操作具有对数级的时间复杂性。std::set 通常是基于红黑树实现的。std::set 的声明如下：

```
template<
    class Key,
    class Compare = std::less<Key>,
    class Allocator = std::allocator<Key>
> class set;

namespace pmr{
    template<
        class Key,
```

```
        class Compare = std::less<Key>>
    using set = std::set<Key, Compare, std::pmr::
    polymorphic_allocator<Key>>;
}
```

下面的示例首先创建了 std::set 对象,然后在 std::set 中插入元素。

【例 5-10】

```cpp
#include <iostream>
#include <iterator>
#include <set>
using namespace std;

int main(){
    set<int, greater<int>> s1;              //std::set 中的元素类型为整型,按降序排列(默认是升序)
    s1.insert(40);
    s1.insert(30);
    s1.insert(60);
    s1.insert(20);
    s1.insert(50);
    s1.insert(50);                          //std::set 中已经有值为 50 的元素,因此值为 50 的元素加不进去
    s1.insert(10);
    set<int, greater<int> >::iterator itr;
    for (itr = s1.begin(); itr != s1.end(); itr++) {
        cout << *itr << " ";
    }
    cout << endl;

    set<int> s2(s1.begin(), s1.end());
    s2.erase(s2.begin(), s2.find(30));      //删除 s2 中值小于或等于 30 的所有元素

    for (itr = s2.begin(); itr != s2.end(); itr++) {
        cout << *itr << " ";
    }

    int num;
    num = s2.erase(50);
    cout << "\ns2.erase(50) : " << num << " removed\n";
    //输出 s1 的上边界和下边界为 40 的元素
    cout << "s1.lower_bound(40) : "
        << *s1.lower_bound(40) << endl;
    cout << "s1.upper_bound(40) : "
        << *s1.upper_bound(40) << endl;
    return 0;
}
```

5.2.2 std::multiset

std::multiset 是一种关联容器,其中包含一组 key 类型的排序对象。与 std::set 不同,std::multiset

允许多个具有等效值的 key。排序可以是升序或者降序。在 std::multiset 中搜索、插入和删除元素的操作具有对数级的时间复杂性。std::multiset 的声明如下：

```
template<
    class Key, class Compare = std::less<Key>, class Allocator = std::allocator<Key>> class multiset;

namespace pmr{
    template<class Key, class Compare = std::less<Key>>
    using multiset = std::multiset<Key, Compare, std::pmr::polymorphic_allocator<Key>>;
}
```

下面的示例创建了 std::multiset 对象并展示了其特性。

【例 5-11】

```
#include <iostream>
#include <set>

using namespace std;

int main(){
    multiset<int> s;                //创建一个 std::multiset 对象，元素以默认的升序排序
    s.insert(15);
    s.insert(15);
    s.insert(15);
    cout << s.count(15) << endl;    //输出 3
    s.erase(s.find(15));            //只删除一个值为 15 的元素
    cout << s.count(15) << endl;    //输出 2
    s.erase(15);                    //删除所有值为 15 的元素
    cout << s.count(15) << endl;    //输出为 0

    return 0;
}
```

在上面的代码中，s.find(15)的作用是找到值为 15 的元素，找到后迭代器定位在该元素，因此，s.erase(s.find(15))的作用是只删除定位器处的元素。

5.2.3　std::map

std::map 是一种以映射方式存储元素的关联容器，其中的每个元素都有一个 key 和一个映射值。任何两个映射值都不能具有相同的 key。std::map 是一类模板，它在<map>头文件中定义，语法如下：

```
template<class Key, class T, class Compare = std::less<Key>, class Allocator = std::allocator<std::pair<const Key, T>>
> class map;

namespace pmr{
    template<class Key, class T, class Compare = std::less<Key>>
    using map = std::map<Key, T, Compare, std::pmr
```

```
            ::polymorphic_allocator<std::pair<const Key, T>>>;
}
```

下面我们看看如何创建 std::map 对象并进行初始化。

【例 5-12】

```cpp
#include <iostream>
#include <map>
#include <string>
#include <string_view>
using namespace std;

void print(string_view s, const map<string,int>&m){
    cout << s;
    for (const auto& [key, value] : m)
        cout << '[' << key << "] = " << value << "; ";
    cout << '\n';
}

int main(){
    map<string,int> m{{"C++",30},{"JAVA",20},{"Python",10}};
    print("1. Initial map: ", m);
    m["C++"] = 35;
    m["C#"] = 25;
    print("2. Updated map: ", m);

    m["JavaScript"] = 25;
    print("3. Updated map: ", m);

    m.erase("C#");
    print("4. After erase: ", m);

    m.clear();
    cout <<boolalpha << "5. Map is empty: " << m.empty() << '\n';
    return 0;
}
```

需要注意的是：在使用 std::map::erase 时要特别小心，下面的代码是有问题的。

```cpp
map<string,string> mapTest;
typedef map<string, string>::iterator ITER;
for(ITER iter = mapTest.begin();iter != mapTest.end(); ++iter){
    mapTest.erase(iter);
}
```

上面的代码会导致程序的不可知行为，原因是 std::map 是一种关联容器。对于关联容器来说，如果一个元素已经被删除，那么其对应的迭代器就失效了，不能再使用。正确的写法是：

```cpp
for(ITER iter = mapTest.begin();iter != mapTest.end();){
    mapTest.erase(iter++);
}
```

当 std::map 中的元素是指针对象时，需要手动删除 std::map 中的指针对象，否则会造成内存泄漏。使用智能指针时，无须手动删除 std::map 中的指针对象。例如：

```
map<string, int*> mapTest;
typedef map<string,int*>::iterator ITER;
for(ITER iter = mapTest.begin(); iter != mapTest.end();){
    if(iter->second)
        delete iter->second;
    mapTest.erase(iter++);
}
```

5.2.4 std::multimap

std::multimap 是一个关联容器，它包含排序后的键-值对的列表，同时允许具有相同键的多个条目。排序可以是升序或者降序。在 std::multimap 中搜索、插入和删除元素的操作具有对数级的时间复杂性。std::multimap 的声明如下：

```
template<class Key, class T, class Compare = std::less<Key>, class Allocator = std::allocator<std::pair<const Key, T>>
> class multimap;

namespace pmr {
template<class Key, class T, class Compare = std::less<Key>>
    using multimap = std::multimap<Key, T, Compare, std::pmr::polymorphic_allocator<std::pair<const Key,T>>>;
}
```

下面的示例给出了创建 std::multimap 对象的代码。

【例 5-13】

```
#include <iostream>
#include <map>
int main (){
    std::multimap<char,int> mmap;
    mmap.insert(std::make_pair('A',12));
    mmap.insert(std::make_pair('B',24));
    mmap.insert(std::make_pair('B',36));
    mmap.insert(std::make_pair('C',58));
    mmap.insert(std::make_pair('C',68));

    std::multimap<char,int>::iterator it;
    for (it=mmap.begin(); it!=mmap.end(); ++it)
        std::cout<<(*it).first<<" => "<<(*it).second<< '\n';
    return 0;
}
```

5.3 无序关联容器

无序关联容器是容器概念的又一次改进。与关联容器一样，无序关联容器将值与 key 相关联，并使用 key 查找值。无序关联容器和关联容器的根本区别在于，关联容器基于树结构，无序关联容器基于另一种形式的数据结构，即哈希表。哈希表由参数 key、hash（一个 hash 函数对象，在 key 上充当哈希函数）和 pred（一个评估 key 之间等价性的二元谓词）组成。std::unordered_map 和 std::unordered_multimap 也具有与 key 关联的映射类型 T。

无序关联容器中的元素被组织到存储桶中，具有相同哈希值的 key 最终位于同一存储桶中。当容器的大小增加时，存储桶的数量会增加，以将每个存储桶中的平均元素数保持在特定值以下。

5.3.1　std::unordered_set

std::unordered_set 是一种无序关联容器，其中包含一组 key 类型的唯一对象。在 std::unordered_set 中搜索、插入和删除元素的操作具有平均常数时间复杂度。std::unordered_set 是使用哈希表实现的，其中 key 是哈希表的哈希索引。std::unordered_set 的声明如下：

```cpp
template<
    class Key,
    class Hash = std::hash<Key>,
    class KeyEqual = std::equal_to<Key>,
    class Allocator = std::allocator<Key>
> class unordered_set;

namespace pmr {
    template<
    class Key,
    class Hash = std::hash<Key>,
    class Pred = std::equal_to<Key>
    > using unordered_set = std::unordered_set<Key, Hash, Pred,
    std::pmr::polymorphic_allocator<Key>>;
}
```

在 std::unordered_set 内部，元素不按任何特定顺序排列，而是组织到存储桶中。将元素放入哪个存储桶中完全取决于该元素的哈希值。这允许快速访问某个元素，因为一旦计算出某个元素的哈希值，就指明了元素被放置的存储桶。std::unordered_set 中的元素不能被修改（即使是非常量迭代器），因为修改可能会改变元素的哈希值并导致 std::unordered_set 崩溃。

下面的代码展示了 std::unordered_set 中的插入、查询、删除和遍历元素等操作。

【例 5-14】

```cpp
#include <iostream>
#include <unordered_set>
#include <string>

using namespace std;

int main(){
    unordered_set<string> s;
    s.insert("C++");
    s.insert("Java");
    s.insert("C#");
```

```cpp
    s.insert("Python");
    s.insert("JavaScript");

    string key = "Python";
    if (s.find(key) == s.end())
        cout << key << " not found" << endl;
    else
        cout << key  << "found " << endl;

    key = "C++";
    unordered_set<string>::iterator iter;
    if(iter = s.find(key); iter != s.end())
        s.erase(iter);

    for (iter = s.begin(); iter != s.end(); iter++)
        cout << (*iter) << endl;
    return 0;
}
```

5.3.2　std::unordered_map

　　std::unordered_map 是一种无序关联容器，其中包含具有唯一 key 的键-值对。元素的搜索、插入和删除操作具有平均常数时间复杂度。在 std::unordered_map 内部，元素不按任何特定的顺序排列，而是组织到存储桶中。将元素放入哪个存储桶完全取决于该元素的 key 的哈希值。具有相同哈希值的 key 出现在同一个存储桶中。这允许快速访问某个元素，因为一旦计算出某个元素的 key 的哈希值，就指明了该元素所在的确切存储桶。std::unordered_map 的声明如下：

```cpp
template<
    class Key,
    class T,
    class Hash = std::hash<Key>,
    class KeyEqual = std::equal_to<Key>,
    class Allocator = std::allocator<std::pair<const Key, T>>
> class unordered_map;

namespace pmr {
    template<
    class Key,
    class T,
    class Hash = std::hash<Key>,
    class KeyEqual = std::equal_to<Key>
    >
    using unordered_map = std::unordered_map<Key, T, Hash, KeyEqual,
                    std::pmr::polymorphic_allocator<std::pair<const Key, T>>>;
}
```

　　下面我们看看如何创建 std::unordered_map 对象并进行初始化，以及如何插入、删除、查询和遍历元素。

【例 5-15】

```cpp
#include <iostream>
#include <unordered_map>
using namespace std;

int main() {
    unordered_map<string, double> fruitPrices ={
        {"Apple", 3.25},
        {"Orange", 2.58},
        {"Pineapple", 3.75}
    };

    fruitPrices["Cherry"] = 5.34;
    fruitPrices["Watermelon"] = 1.46;
    fruitPrices["Pear"] = 3.73;
    fruitPrices["Banana"] = 2.30;
    fruitPrices["Strawberry"] = 2.88;

    fruitPrices.insert(make_pair("Mango", 3.46));

    string key = "Lemon";
    if (fruitPrices.find(key) != fruitPrices.end())
        fruitPrices.erase(key);
    else
        cout << key << "not found" <<    "\n";

    unordered_map<string, double>::iterator iter;
    for(iter=fruitPrices.begin();iter!=fruitPrices.end();itr++){
        cout << iter->first << " " << itr->second << endl;
    }
    return 0;
}
```

需要注意的是：在使用 std::unordered_map::erase 时要特别小心，需要采用如下的方式。

【例 5-16】

```cpp
#include <unordered_map>
#include <iostream>

int main(){
    std::unordered_map<int, std::string> c = {
        {1, "one"}, {2, "two"}, {3, "three"},
        {4, "four"}, {5, "five"}, {6, "six"}
    };
    //删除所有奇数的元素
    for (auto it = c.begin(); it != c.end();){
        if ((it->first%2)!=0)
            it = c.erase(it);
        else
```

```
            ++it;
        }
    for (auto& p : c)
        std::cout << p.second << '\n';
    return 0;
}
```

下面的示例展示 std::unordered_map 在词频统计方面的应用。

【例 5-17】

```
#include <iostream>
#include <fstream>
#include <unordered_map>
#include <sstream>
#include <cctype>

int main() {
    std::string filename = "example.txt";        //读取文件中的单词，进行词频统计
    std::ifstream file(filename);
    std::unordered_map<std::string, int> wordCount;
    if (!file.is_open()) {
        std::cerr <<"Failed to open file: "<<filename<<std::endl;
        return 1;
    }
    std::string word;
    while (file >> word) {
        //删除单词中的标点符号
        word.erase(std::remove_if(word.begin(), word.end(), ::ispunct), word.end());
        //将字母转换为小写
        std::transform(word.begin(),word.end(),word.begin(), ::tolower);
        //如果单词不为空，则更新计数
        if (!word.empty()) {
            wordCount[word]++;
        }
    }
    file.close();                                //输出单词及其频率
    for (const auto &pair : wordCount){
        std::cout << pair.first <<":"<<pair.second<< std::endl;
    }
    return 0;
}
```

5.3.3　std::unordered_multiset

　　std::unordered_multiset 是一个关联容器，其中包含一组可能非唯一 key 类型的对象。搜索、插入和删除等操作具有平均常数时间复杂度。在 std::unordered_multiset 内部，元素不按任何特定顺序排列，而是组织到桶中。将元素放入哪个桶中完全取决于其 key 的哈希值。这允许快速访问单个元素，因为一旦计算出哈希值，就指明了该元素放置的存储桶。

std::unordered_multiset 的声明如下：

```
template<
    class Key,
    class Hash = std::hash<Key>,
    class KeyEqual = std::equal_to<Key>,
    class Allocator = std::allocator<Key>
> class unordered_multiset;

namespace pmr {
    template<
    class Key,
    class Hash = std::hash<Key>,
    class Pred = std::equal_to<Key>
    >
    using unordered_multiset=std::unordered_multiset<Key,Hash,
    Pred,std::pmr::polymorphic_allocator<Key>>;
}
```

下面的示例创建了 std::unordered_multiset 对象并进行了初始化，实现了元素的插入、统计、查询、删除、清空、遍历等操作。

【例 5-18】

```cpp
#include <iostream>
#include <string>
#include <unordered_set>

int main (){
    std::unordered_multiset<std::string> first;
    std::unordered_multiset<std::string> second({"red","green","blue"});
    std::unordered_multiset<std::string> third({"red","yellow","blue"});
    std::unordered_multiset<std::string> fourth(second);
    std::unordered_multiset<std::string>fifth(third.begin(), third.end());
    first = {"red","yellow","blue"};
    first.insert("black");
    first.insert("white");

    string val = "blue";
    if(first.find(val) != first.end())
    cout << "unordered multiset first contains: " << val;
    else
    cout << "unordered multiset first doesn't contains: " << val;

    val = "black";
    int cnt = first.count(val);
    cout << val << " appears " << cnt << " times in first\n";

    val = "blue";
    first.erase(val);
```

```cpp
    unordered_multiset<std::string>::iterator iter;
    for(iter = first.begin(); iter!= first.end();)
    std::cout << *iter << "   ";
    std::cout << std::endl;
    first.clear();
    return 0;
}
```

5.3.4　std::unordered_multisetmap

　　std::unordered_multimap 是一种无序关联容器，它支持等效 key，即它的每个 key 可能包含多个副本。std::unordered_multimap 支持正向迭代器，搜索、插入和删除等操作具有平均常数时间复杂度。在 std::unordered_multimap 内部，元素不按任何特定顺序排列，而是组织到存储桶中。将元素放入哪个存储桶完全取决于元素的哈希值，这允许快速访问某个元素，因为一旦计算出该元素的哈希值，就指明了该元素所在的存储桶。

　　std::unordered_multimap 的声明如下：

```cpp
template<
    class Key,
    class T,
    class Hash = std::hash<Key>,
    class KeyEqual = std::equal_to<Key>,
    class Allocator = std::allocator<std::pair<const Key, T>>
> class unordered_multimap;

namespace pmr {
    template<
    class Key,
    class T,
    class Hash = std::hash<Key>,
    class Pred = std::equal_to<Key>
    >
    using unordered_multimap=std::unordered_multimap<Key, T, Hash, Pred, std::pmr::polymorphic_allocator<std::pair<const Key, T>>>;
}
```

　　下面的示例创建了 std::unordered_multimap 对象并进行了初始化，实现了元素的插入、查询、删除、交换、遍历等操作。

【例 5-19】

```cpp
#include <iostream>
#include <unordered_map>
int main (){
    std::unordered_multimap<std::string,std::string> fruits ={
        {"strawberry","red"},
        {"banana","yellow"},
        {"orange","orange"}
    };
```

```cpp
    std::pair<std::string,std::string> pear ("pear","green");
    fruits.insert(pear);
    first.insert ({{"apple","red"},{"apple","green"}});
    fruits.emplace({"lemon","yellow"});
    fruits.erase ( fruits.begin() );
    fruits.erase ("apple");
    fruits.erase ( fruits.find("orange"), fruits.end() );

    for (auto& x: {"orange","lemon","strawberry"}) {
        std::cout << x << ": " << fruits.count(x) << " entries.\n";
    }

    for ( auto it = fruits.begin(); it!= fruits.end(); ++it )
    std::cout << " " << it->first << ":" << it->second;

    std::cout << std::endl;

    std::unordered_multimap<std::string,std::string> g;
    fruits.swap(g);

    g.clear();
    return 0;
}
```

5.4 容器适配器

容器适配器是 C++标准模板库的一部分，它提供了一种修改或调整现有容器类以满足特定需求或要求的方法。在 C++中，容器适配器是在顺序容器（如 std::deque、std::vector 或 std::list）上创建的专用接口，用来限制或重新封装底层容器的访问接口，提供特定数据结构的抽象行为。这样做是为了避免为容器定义一个全新的接口。到 C++20 为止，容器适配器有三种：std::stack、std::queue 和 std::priority_queue。

5.4.1 std::stack

std::stack 是一种类模板的容器适配器，具有堆栈 LIFO（后进先出）的功能。std::stack 充当基础容器的包装器，仅提供一组特定的函数。堆栈从容器的尾部（称为堆栈顶部）压进和弹出元素。std::stack 的声明如下：

```cpp
template< class T, class Container = std::deque<T>>
class stack;
```

下面的示例实现了一个简单的与堆栈相关的操作。

【例 5-20】

```cpp
#include <iostream>
```

```cpp
#include <stack>
int main (){
    std::stack<std::string> s;
    s.push("apple");
    s.push("orange");
    s.push("banana");
    s.push("pineapple");

    std::cout << "Popping out elements...";
    while(!s.empty()){
        std::cout << s.top() << endl;
        s.pop();
    }
    return 0;
}
```

5.4.2　std::queue

std::queue 是一种类模板的容器适配器，它提供队列 FIFO（先进先出）的功能和数据结构。std::queue 充当基础容器的包装器，仅提供一组特定的函数，它将元素从队列的尾部推进，并从队列的头部弹出。std::queue 的声明如下：

```
template< class T, class Container=std::deque<T>>
class queue;
```

下面的示例实现了一个简单的与队列相关的操作。

【例 5-21】

```cpp
#include <cassert>
#include <iostream>
#include <queue>
int main(){
    std::queue<int> q;
    for(int i=0; i<10; i++)
    q.push(i);

    assert(q.front() == 0);
    assert(q.back() == 9);
    assert(q.size() == 10);

    q.pop();
    q.pop();
    assert(q.size() == 8);

    for (; !q.empty();){
        std::cout << q.front() << ' ';
        q.pop();
    }
    std::cout << '\n';
```

```
        assert(q.size() == 0);
}
```

5.4.3 std::priority_queue

std::priority_queue 是一种容器适配器,在默认情况下,它提供最大元素的恒定时间查找,但插入和提取元素的时间复杂度是对数级的。std::priority_queue 可以使用比较功能来更改元素的排序,例如使用 std::greater<T>会导致最小的元素出现在 top()。std::priority_queue 的使用类似于在某个随机存取的容器中管理队列,其好处是不会意外地使队列失效。

```
template<
    class T,
    class Container = std::vector<T>,

    class Compare = std::less<typename Container::value_type>
> class priority_queue;
```

下面的示例实现了一个简单的与优先队列相关的操作。

【例 5-22】

```
#include <iostream>
#include <queue>
int main (){
    std::priority_queue<std::string> pq;
    pq.push("red");
    pq.push("green");
    pq.push("blue");
    pq.push("black");
    pq.push("white");

    std::cout << "Popping out elements...";
    while (!pq.empty()){
        std::cout << pq.top() << '\n';
        pq.pop();
    }
    return 0;
}
```

下面我们用 std::priority_queue 实现一个简单的股票交易系统的原型,它包括下单、改单、撤单、促成交易等功能。

【例 5-23】

```
#include <iostream>
#include <vector>
#include <queue>
#include <algorithm>

//订单结构体
struct Order {
```

```cpp
    int orderId;
    std::string stockSymbol;
    int quantity;
    double price;
    bool isBuy;
    //构造函数
    Order(int id, const std::string& symbol, int qty, double prc)
        : orderId(id), stockSymbol(symbol), quantity(qty), price(prc), isBuy(buy) {}

    //重载比较运算符，用于优先队列
    bool operator<(const Order& other) const {
        if (isBuy != other.isBuy) {
            return false;
        }
        if (isBuy) {
            return price < other.price;          //买入订单，价格高的优先
        }
        return price > other.price;              //卖出订单，价格低的优先
    }
};

//交易系统类
class TradingSystem {
private:
    std::priority_queue<Order> buyOrders;        //买单队列
    std::priority_queue<Order> sellOrders;       //卖单队列
    int nextOrderId = 1;

public:
    //下单
    int placeOrder(const std::string& symbol, int quantity, double price, bool isBuy) {
        Order newOrder(nextOrderId++, symbol, quantity, price, isBuy);
        if (isBuy) {
            buyOrders.push(newOrder);            //新买单
        } else {
            sellOrders.push(newOrder);           //新卖单
        }
        return newOrder.orderId;
    }

    //改单
    bool modifyOrder(int orderId, int newQuantity, double newPrice, bool isBuy ) {
        std::vector<Order> temp;
        bool found = false;
        if (buyOrders.empty() && sellOrders.empty()) {
            return false;
        }
        auto& orders = isBuy ? buyOrders : sellOrders;
        while (!orders.empty()) {
```

```cpp
            Order current = orders.top();
            orders.pop();
            if (current.orderId == orderId && current.isActive) {
                current.quantity = newQuantity;
                current.price = newPrice;
                found = true;
            }
            temp.push_back(current);
        }
        for (const auto& order : temp) {
            if (isBuy) {
                buyOrders.push(order);
            } else {
                sellOrders.push(order);
            }
        }
        return found;
    }

    //撤单
    bool cancelOrder(int orderId, bool isBuy) {
        std::vector<Order> temp;
        bool found = false;
        if (buyOrders.empty() && sellOrders.empty()) {
            return false;
        }
        auto& orders = isBuy ? buyOrders : sellOrders;
        while (!orders.empty()) {
            Order current = orders.top();
            orders.pop();
            if (current.orderId == orderId ) {
                found = true;
            }
            else
                temp.push_back(current);
        }
        for (const auto& order : temp) {
            if (isBuy) {
                buyOrders.push(order);
            } else {
                sellOrders.push(order);
            }
        }
        return found;
    }

    //成交
    void executeTrades() {
        while (1){
```

```cpp
            if (buyOrders.empty() || sellOrders.empty())
                return;
            Order buy = buyOrders.top();
            Order sell = sellOrders.top();
            if (buy.stockSymbol != sell.stockSymbol) {
                continue;
            }

            if (buy.price >= sell.price) {
                int tradeQuantity = std::min(buy.quantity, sell.quantity);
                std::cout << "交易成功: 股票代码 " << buy.stockSymbol
                          << ", 数量 " << tradeQuantity
                          << ", 价格 " << sell.price << std::endl;

                buy.quantity -= tradeQuantity;
                sell.quantity -= tradeQuantity;

                buyOrders.pop();
                sellOrders.pop();

                if (buy.quantity > 0) {
                    buyOrders.push(buy);
                }
                if (sell.quantity > 0) {
                    sellOrders.push(sell);
                }
            } else {
                continue;
            }
        }
    }
};

int main() {
    TradingSystem tradingSystem;

    //下单
    int orderId1 = tradingSystem.placeOrder("ABC", 100, 10.0, true);
    int orderId2 = tradingSystem.placeOrder("ABC", 200, 11.0, true);
    int orderId3 = tradingSystem.placeOrder("ABC", 150, 9.5, false);

    //改单
    tradingSystem.modifyOrder(orderId1, 150, 10.5, true);

    //促成交易
    tradingSystem.executeTrades();

    //撤单
    tradingSystem.cancelOrder(orderId2, true);
```

```
        return 0;
}
```

5.5 分配器与迭代器

STL 的容器都有一个类型参数 allocator，在默认情况下是 std::allocator。这说明我们可以定制容器的分配器，从而管理对象的内存分配。同时每一种容器都有成员类型 Iterator（迭代器），有的容器还有双向迭代器。

5.5.1 std::allocator

std::allocator 是 STL 容器的默认分配器，负责封装内存管理的对象。std::allocator 仅使用运算符 new 和 delete 来获取和释放内存。std::allocator 的声明如下：

```
template< class T >
struct allocator;
```

【例 5-24】

```
#include <iostream>
#include <memory>
using namespace std;

int main(){
    allocator<int> myAllocator;
    int* arr = myAllocator.allocate(5);        //申请分配 5 个 int 类型的数组元素存储单元
    arr[0] = 50;
    arr[1] = 60;
    arr[2] = 70;
    arr[3] = 80;
    arr[4] = 90;
    for(int i=0; i<5; i++)
        cout << arr[i] << ' ' << endl;

    myAllocator.deallocate(arr, 5);            //释放内存
    return 0;
}
```

std::allocator 的两个主要成员函数是 allocate()和 deallocate()，其他成员函数在 C++20 中皆已被弃用。当使用 std::allocator 作为容器的分配器时，并不直接调用其成员函数 allocate()和 deallocate()来分配对象的内存和释放内存，而是由容器来隐式调用成员函数的。

5.5.2 迭代器

迭代器在连接容器与算法时起关键作用。迭代器是一种检查容器内元素并遍历元素的数据类

型，通常可以简单地将迭代器理解为指针，用于对容器内元素进行访问，但不同的容器有不同的迭代器。

迭代器的种类有输入迭代器（input iterator）、输出迭代器（output iterator）、前向迭代器（forward iterator）、双向迭代器（bidirectional iterator）、随机访问迭代器（random-access iterator）。

使用迭代器遍历容器的元素时，通常使用容器的两个成员函数：begin()和end()。begin()指向容器第一个元素，但end()不是指向容器最后一个元素的，而是指向容器最后一个元素的下一个位置。当迭代器到达end()指向的位置时，说明迭代器已经遍历了容器的所有元素。

在STL的容器中，除std::stack和std::queue不支持迭代器外，其他容器支持的迭代器类型如下：
- std::array 支持随机访问迭代器。
- std::vector 支持随机访问迭代器。
- std::deque 支持随机访问迭代器。
- std::list 支持双向迭代器。
- std::forward_list 支持前向迭代器。
- std::set 和 std::multiset 支持双向迭代器。
- std::map 和 std::multimap 支持双向迭代器。

迭代器访问操作属性如表 5.5-1 所示。

表 5.5-1 迭代器访问操作属性

迭代器	属性				
	访问	读	写	分配	比较
输入迭代器	->	=*iter		++	==、!=
输出迭代器			*iter=	++	
前向迭代器	->	=*iter	*iter=	++	==、!=
双向迭代器		=*iter	*iter=	++、--	==、!=
随机访问迭代器	->、[]	=*iter	*iter=	++、--、+、-、+=、-=	==、!=、>=、<=、>、<

5.5.3 迭代器失效

当迭代器所指向的容器内部的元素发生变化时，即当元素从一个位置移动到另一个位置，且迭代器仍然指向旧的无效位置，则称为迭代器失效。在C++中使用迭代器时应该小心。当我们使用迭代器遍历容器中的元素时，迭代器可能会失效。这可能是由于容器的形状发生了变化或者大小发生了变化。例如，std::vector 有动态扩展的功能，当其容量不足时，std::vector 就会进行动态扩容，动态扩容不是在原来的空间后面追加空间，而是在寻找新的更大的空间，把原来的元素复制过去，导致 std::vector 中元素的存储位置发生了变化，但迭代器还是指向原来的位置，因此每次进行动态扩容后原来的迭代器都会失效。

5.6 本章小结

本章主要介绍 C++标准模板库的容器类。所有的容器都被设计为类模板，所以支持存储各种数据类型。容器分为顺序容器、关联容器、无序关联容器、容器适配器。每一种容器都定义了一组操作容器中元素的函数。

顺序容器的特点是以线性排列方式存储相同类型的数据。std::array 和 std::vector 是在连续的内存空间中存储数据的，因此可以通过索引来访问元素。std::deque 是分段连续存储数据的，也可以通过索引来访问元素。而 std::list 和 std::forward_list 的数据存储在非连续的内存空间中。std::list 是通过双向链表实现的，std::forward_list 是通过单向链表实现的，它们的遍历速度慢，但插入速度快。std::list 和 std::forward_list 不可以通过索引来访问元素。

当使用顺序容器存储指针对象时，需要手动删除容器内所有的指针对象，否则会造成内存泄漏。使用智能指针时，无须手动删除容器中的指针对象。

关联容器是一个有序容器，它提供基于键的对象快速查找。std::set 和 std::map 是支持唯一键的关联容器，即一个键最多包含一个元素；而 std::multiset 和 std::multimap 支持等效键，即一个键可以包含多个元素。关联容器的默认排序是升序，如果要降序排列元素，则需要在声明时使用 std::greater()比较函数。

对于关联容器来说，使用其成员函数 erase()时要特别小心。这是因为，如果某个元素已经被删除了，那么其对应的迭代器就会失效，不能再使用。另外，如果关联容器中存储的是指针对象，需要写一个循环删除这些指针对象。

无序关联容器是容器概念的又一次改进。关联容器是基于树结构实现的，无序关联容器是基于哈希表实现的，因此无序关联容器是具有相对快速且高效搜索算法的容器。对于无序关联容器来说，使用其成员函数 erase()时要特别小心。另外，如果无序关联容器中存储的是指针对象，需要写一个循环来删除这些指针对象。

容器适配器是在顺序容器（如 std::deque、std::vector 或 std::list）上创建的专用接口，用来限制或重新封装底层容器的访问接口，提供特定数据结构的抽象行为。这样做是为了避免为容器定义一个全新的接口。

第 6 章 STL 函数

STL 提供了不同类型的算法，这些算法可以在迭代器的帮助下在任何容器上实现。因此，我们不必自己定义复杂的算法，只需使用 STL 算法库<algorithm>提供的内置算法函数即可。STL 中的算法都是函数模板。使用 STL 的算法可以节省时间、精力和代码，并且非常可靠。

算法库<algorithm>中的算法主要包括排序算法、搜索算法、非更改顺序算法、更改顺序算法、分割算法、合并算法、堆算法、最大最小值算法、数值算法。

使用 STL 算法库中的算法函数时，需要在代码中添加"#include <algorithm>"。本章在介绍算法函数后，还介绍了头文件<functional>中的函数对象、<utility>中的常用函数，以及回调函数。

6.1 算法函数

6.1.1 排序算法

STL 算法库中涉及排序的算法有 std::sort、std::is_sorted、std::stable_sort、std::partial_sort、std::nth_element。限于篇幅，本节只介绍 std::sort。std::sort 有以下两种语法格式：

```
template< class RandomIt >
void sort( RandomIt first, RandomIt last );

template< class RandomIt, class Compare >
void sort( RandomIt first, RandomIt last, Compare comp );
```

这里 first 和 last 是用于排序的容器元素的起始位置和最后位置。std::sort 的排序方式有两种：第一种是默认的排序算法，即升序排序；第二种是自定义排序，需要指定比较函数。例如：

【例 6-1】

```cpp
#include <algorithm>
#include <array>
#include <functional>
#include <iostream>
#include <string_view>

int main(){
    std::array<int, 10> a{26, 17, 23, 18, 36, 64, 51, 48, 37, 43};
    auto print_output = [&a](std::string_view const rem){
        for (auto i : a)
```

```cpp
            std::cout << i << ' ';
        std::cout << ": " << rem << '\n';
    };

    print_output("使用默认的比较函数排序");
    std::sort(a.begin(), a.end());                    //默认为 std::less<int>()

    print_output("使用指定的 STL 比较函数排序");
    std::sort(a.begin(), a.end(), std::greater<int>());

    struct compare{
        bool operator()(int x, int y) const { return x < y; }
    };

    compare Less;
    print("使用定制的比较函数排序");
    std::sort(a.begin(), a.end(), Less);

    print("使用 lambda 函数排序");
    std::sort(a.begin(), a.end(), [](int x, int y){
        return x > y;
    });

    return 0;
}
```

6.1.2 搜索算法

STL 算法库中与搜索相关的算法有 std::lower_bound、std::upper_bound、std::equal_range、std::binary_search。限于篇幅，本节只介绍 std::binary_search。std::binary_search 的语法如下：

```cpp
template< class ForwardIt, class T >
bool binary_search( ForwardIt first, ForwardIt last, const T& value );

template< class ForwardIt, class T, class Compare >
bool binary_search( ForwardIt first, ForwardIt last, const T& value, Compare comp );
```

binary_search(first, last, value)：如果在给定的范围内，即(first,last)，存在元素 a，且满足条件 (!(a < value) && !(value < a))，则返回 true。换句话说，运算符用于检查两个元素的相等性。

binary_search(first, last, value, compare_function)：如果在给定的范围内，即(first,last)，存在元素 a，且满足比较函数 compare_function，则返回 true。

【例 6-2】

```cpp
#include <iostream>
#include <algorithm>
#include <vector>
using namespace std;
```

```cpp
bool compare_string (const string a,const string b){
    return (a == b);
}

int main () {
    int inputs[] = {24,29,18,26,32,38,29,41};
    vector<int> v(inputs, inputs + 8);

    cout<<binary_search(v.begin() , v.end() , 26 );
    cout<<binary_search(v.begin() , v.end() , 53 );

    vector<string> s;
    s.push_back("hello");
    s.push_back("thanks");
    s.push_back("C++");
    s.push_back("cat");

    //在 s 中搜索字符串"dog"
    cout << binary_search(s.begin(),s.end(),"dog",compare_string);
    return 0;
}
```

6.1.3 非更改顺序算法

非更改顺序算法是指算法在执行过程中，不对容器内的元素值或顺序进行修改。这类算法主要有 std::count、std::count_if、std::equal、std::mismatch、std::search、std::search_n。

1. std::count 和 std::count_if

std::count 和 std::count_if 返回指定范围内满足给定条件的元素个数，语法为：

```cpp
template< class InputIt, class T >
typename std::iterator_traits<InputIt>::difference_type
    count(InputIt first, InputIt last, const T& value);


template< class InputIt, class UnaryPred >
typename std::iterator_traits<InputIt>::difference_type
    count_if(InputIt first, InputIt last, UnaryPred p);
```

count(first, last, value)：返回由迭代器(first,last）定义的范围内元素的数量，数量等于 value。

count_if(first, last, p)：返回由迭代器(first,last）定义的范围内元素的数量，若元素满足一元谓词 p 则被统计。

【例 6-3】

```cpp
#include <iostream>
#include <algorithm>
#include <vector>
using namespace std;

int main (){
```

```cpp
    int a[] = {6,15,5,8,19,15,12,6,8,15,12,8,12,15,6};
    int count_8 = count(a, a+15, 8);                    //count_8:3
    vector<int> v(a, a+15);
    int count_15 = count(v.begin(),v.end(),15);         //count_15:4

    int count = count_if(v.begin(),v.end(),[](int i){ return i%5==0; });
    cout << "numbers divisible by 5: " << count << '\n';
    return 0;
}
```

2. std::equal

std::equal 用于比较两个范围内的元素，如果一个范围内的所有元素都与另一个范围内的相应元素都相等，则返回 true，否则返回 false。std:: equal 有两个变体：

```cpp
template< class InputIt1, class InputIt2 >
bool equal( InputIt1 first1, InputIt1 last1, InputIt2 first2 );


template< class InputIt1, class InputIt2, class BinaryPred >
bool equal( InputIt1 first1, InputIt1 last1, InputIt2 first2, InputIt2 last2, BinaryPred p );
```

equal(first1, last1, first2)：将[first1,last1)内的元素与起始位置为 first2 的 last1－first1 个元素进行相等性比较，如果所有的元素都相等，则返回 true；否则返回 false。

equal(first1, last1, first2, cmp_function)：这里 cmp_function 用于决定如何检查两个元素的相等性，它对于字符串和对象等非数字元素很有用。

【例 6-4】

```cpp
#include <iostream>
#include <algorithm>
#include <vector>
using namespace std;

bool cmp_string(const string i, const string j){
    return (i == j);
}

int main(){
    int a[] = { 1,2,3,4,5,6,7,8,9};
    int b[] = { -1,2,1,2,3,4,5,6,7,8,9};

    vector<int>   v1(a , a+9 );
    vector<int>   v2(b , b+11 );

    cout << equal(v1.begin(), v1.end(), v2.begin()+2 );

    //使用比较函数
    string s1[] = { "cat" , "dog" , "cow" , "rabbit" };
    string s2[] = { "pig" , "horse" , "duck" , "chicken" };

    cout<<equal( s1 , s1+4 , s2 , cmp_string);
```

```
        return 0;
}
```

3. std::mismatch

std::mismatch 返回一对迭代器,其中第一个迭代器指向第一个容器中的元素,第二个迭代器指向第二个容器中发生不匹配的元素。std::mismatch 有以下两个变体:

```
template< class InputIt1, class InputIt2 >
std::pair<InputIt1, InputIt2>
mismatch(InputIt1 first1, InputIt1 last1,InputIt2 first2);

template< class InputIt1, class InputIt2, class BinaryPred >
std::pair<InputIt1, InputIt2>
mismatch(InputIt1 first1, InputIt1 last1, InputIt2 first2, BinaryPred p );
```

mismatch(first1, last1, first2):这里的 first1 和 last1 用于指定第一个容器的迭代器范围,first2 是第二个容器的迭代器指定的开始比较的位置。在默认情况下,会检查元素是否相等,并返回一对迭代器,给出发生不匹配的元素的位置。

mismatch(first1, last1, first2, p):其工作方式与上述变体相同,其中的二元谓词 p 用于检查元素是否相等。

【例 6-5】

```
#include<iostream>
#include<algorithm>
#include<vector>
using namespace std;

bool cmp_string(const string i , const string j){
    return ( i.size() == j.size() );
}

int main(){
    int array1[] = {1,2,3,4,5,6,7,8};
    int array2[] = {-1,2,1,2,3,4,5,6,7,9,0};

    vector<int> v1(array1 ,array1+8);
    vector<int> v2(array2 ,array2+11);

    //定义一对迭代器
    pair<vector<int>::iterator, vector<int>::iterator> position;
    position = mismatch(v1.begin(), v1.end(), v2.begin()+2) ;

    //使用比较函数
    string s1[] = {"cat" , "dog" , "cow" , "rabbit"};
    string s2[] = {"pig" , "horse" , "duck" , "chicken"};

    pair<string::iterator, string::iterator> position2;
    position2 = mismatch( s1, s1+4, s2, cmp_string);
    return 0;
}
```

4. std::search

std::search 用于在给定范围内搜索给定的序列。std::search 有两个变体：

```
template< class ForwardIt1, class ForwardIt2 >
ForwardIt1 search( ForwardIt1 first1, ForwardIt1 last1, ForwardIt2 first2, ForwardIt2 last2);

template<class ForwardIt1,class ForwardIt2,class BinaryPred>
ForwardIt1 search( ForwardIt1 first1, ForwardIt1 last1, ForwardIt2 first2, ForwardIt2 last2, BinaryPred p);
```

search(first1, last1, first2, last2)：在[first1, last1)内搜索由 first2 和 last2 定义的序列，如果存在匹配项，则返回[first1, last1)内指向匹配项第一个元素的迭代器，否则返回指向 last1 的迭代器。

search(first1, last1, first2, last2, p)：这里 p 用于决定如何检查两个元素的相等性，它对于字符串和对象等非数字元素很有用。

【例 6-6】

```cpp
#include<iostream>
#include<algorithm>
#include<vector>
using namespace std;

int main(){
    int a[] = { 1,2,3,4,5,6,7,8,9};
    int b[] = { 3,4,5,6 };

    vector<int> v1(a, a+9);
    vector<int> v2(b, b+4);

    vector<int>::iterator iter1 ,iter2;

    //现在 iter1 指向 v1 中的第 3 个元素
    iter1 = search(v1.begin(), v1.end(), v2.begin(), v2.end());
    //iter2 指向 v1 的末尾，因为没有匹配项
    iter2 = search(v1.begin()+4, v1.end(), v2.begin(), v2.end());
}
```

5. std::search_n

std::search_n 可在[first, last)内中搜索指定个数相同元素的序列，每个元素都等于给定值。std::search_n 有两个变体：

```
template< class ForwardIt, class Size, class T >
ForwardIt search_n( ForwardIt first, ForwardIt last, Size count, const T& value );

template<class ForwardIt,class Size,class T,class BinaryPre>
ForwardIt search_n( ForwardIt first, ForwardIt last, Size count, const T& value, BinaryPred p);
```

search_n(first, last, count, value)：在[first, last)内搜索 count 个 value 的序列，如果存在匹配项，则返回[first, last)内指向第一个元素的迭代器，否则返回指向 last 的迭代器。

search_n(first, last, count, value, p)：这里 p 用于决定如何检查两个元素的匹配，它对字符串和对象等非数字元素很有用。

【例 6-7】

```cpp
#include<iostream>
#include<algorithm>
#include<vector>
using namespace std;

int main(){
    int i, j;
    vector<int> a = { 3, 1, 5, 3, 8, 5, 3, 3, 2 };
    int b = 3;

    vector<int>::iterator iter;
    iter = std::search_n(a.begin(), a.end(), 2, b);

    if (iter != a.end()) {
        cout << "3 在 a 中连续出现 2 次,位置在: "
            << (iter - a.begin());
    }
    else {
        cout << "3 没有在 a 中连续出现 2 次";
    }
    return 0;
}
```

6.1.4 更改顺序算法

更改顺序算法是指算法在执行过程中,会对容器内元素的值进行修改或者对元素顺序进行修改。STL 算法库中更改顺序算法的函数有 std::copy、std::copy_n、std::fill、std::fill_n、std::rotate、std::transform、std::generate、std::swap、std::swap_ranges、std::shuffle、std::reverse、std::reverse_copy、std::unique、std::unique_copy 等。

1. std::copy 和 std::copy_if

将[first, last)内的元素复制到另一个从 d_first 开始的区间,或者仅复制[first, last)中满足一元谓词 pred(若满足一元谓词 pred 则返回为 true)的元素。这种复制算法是稳定的,可保留被复制元素的相对顺序。std::copy 和 std::copy_if 的语法如下:

```cpp
template< class InputIt, class OutputIt >
OutputIt copy(InputIt first, InputIt last, OutputIt d_first);

template< class InputIt, class OutputIt, class UnaryPred >
OutputIt copy_if( InputIt first, InputIt last, OutputIt d_first, UnaryPred pred );
```

【例 6-8】

```cpp
#include <algorithm>
#include <iostream>
#include <iterator>
#include <numeric>
```

```cpp
#include <vector>

int main(){
    std::vector<int> from{3,5,6,8,2,9,7,1,4};

    std::vector<int> to(from.size());
    std::copy(from.begin(),from.end(),to.begin());

    std::cout << "打印向量数组中的奇数:";
    std::copy_if(to.begin(), to.end(),
                 std::ostream_iterator<int>(std::cout, " "),
                 [](int x) { return x % 2 != 0;});

    std::cout << endl;

    to.clear();
    //将向量 from 中 3 的倍数复制到 to 中
    std::copy_if(from.begin(), from.end(),
                 std::back_inserter(to),
                 [](int x) { return x % 3 == 0; });

    for (const int x : to)
        std::cout << x << ' ';
    std::cout << endl;
    return 0;
}
```

2. std::fill

std::fill 可将给定的值赋值给指定范围的元素，其语法如下：

```cpp
template< class ForwardIt, class T >
void fill( ForwardIt first, ForwardIt last, const T& value);
```

【例 6-9】

```cpp
#include <iostream>
#include <algorithm>
#include <vector>
using namespace std;

int main () {
    vector<int> v(10);
    fill(v.begin(), v.end(), 8);
    fill(v.begin(), v.end()-3, 5);
    //现在 v 的元素为 5、5、5、5、5、5、5、8、8、8
    return 0;
}
```

3. std::generate

std::generate 可为[first, last)内的每个元素分配一个由给定函数对象 g 生成的值，其语法格式如下：

```
template< class ForwardIt, class Generator >
void generate(ForwardIt first, ForwardIt last, Generator g);
```

【例 6-10】

```
#include <iostream>
#include <vector>
#include <algorithm>
using namespace std;

int gen(){
    static int i = 0;
    return ++i;
}

int main(){
    int i;
    vector<int> v(10);
    std::generate(v.begin(), v.end(), gen);
    vector<int>::iterator iter;
    for (iter = v.begin(); iter != v.end(); ++iter){
        cout << *iter << ' ';
    }
    return 0;
}
```

4．std::transform

std::transform 可将给定函数应用于给定输入范围的元素，并将结果存储在从 d_first 开始的范围内。std::transform 有两个变体，语法如下：

```
template< class InputIt, class OutputIt, class UnaryOp >
OutputIt transform( InputIt first1, InputIt last1, OutputIt d_first, UnaryOp unary_op );

template< class InputIt1, class InputIt2, class OutputIt, class BinaryOp >
OutputIt transform( InputIt1 first1, InputIt1 last1, InputIt2 first2, OutputIt d_first, BinaryOp binary_op );
```

在第一种变体中，一元函数对象 unary_op 将应用于[first1, last1)内的每一个元素。在第二种变体中，二元函数对象 binary_op 将应用于元素对上，元素对分别来自[first1, last1)和[first2, last1)的对应元素。

【例 6-11】

```
#include <iostream>
#include <algorithm>
#include <vector>
#include <functional>

int op_square (int i) { return i*i; }

int main (){
    std::vector<int> v1;
```

```cpp
    std::vector<int> v2;

    for (int i=1; i<10; i++)
        v1.push_back (i);                    //v1: 1  2  3  4  5  6  7  8  9
    v2.resize(v1.size());

    std::transform(v1.begin(),v1.end(),v2.begin(), op_square);
    //v2: 1   4   9   16   25   36   49   64   81

    //std::plus 返回两个参数相加的结果
    std::transform(v1.begin(),v1.end(),v2.begin(),v1.begin(), std::plus<int>());

    std::cout << "v1: ";
    std::vector<int>::iterator it=v1.begin();
    for(; it!=v1.end(); ++it)
        std::cout << *it <<' ';

    std::cout << endl;
    return 0;
}
```

5．std::rotate

std::rotate 可对[first, last)内的元素执行左旋转操作。具体来说，std::rotate 可交换[first, last)内的元素，将[first, middle)内的元素放在[middle, last)内的元素之后，同时保留两个区间内元素的顺序不变。std::rotate 的语法如下：

```cpp
template< class ForwardIt >
ForwardIt rotate(ForwardIt first, ForwardIt middle, ForwardIt last);
```

【例 6-12】

```cpp
#include <algorithm>
#include <iostream>
#include <vector>

void print(std::string_view const rem, std::vector<int>& v){
    std::cout << rem;
    vector<int>::iterator iter;
    for (iter = v.begin(); iter != v.end(); ++iter){
        cout << *iter <<' ';
    }
    std::cout << endl;
};

int main(){
    std::vector<int> v{2, 4, 2, 0, 5, 10, 7, 3, 7, 1};
    print("before sort:\t\t", v);            // 2 4 2 0 5 10 7 3 7 1

    //插入排序
    for (auto i = v.begin(); i != v.end(); ++i)
```

```cpp
        std::rotate(std::upper_bound(v.begin(), i, *i), i, i + 1);
    print("after sort:\t\t", v);              // 0 1 2 2 3 4 5 7 7 10

    //简单向左旋转
    std::rotate(v.begin(), v.begin() + 1, v.end());
    print("simple rotate left:\t", v);        // 1 2 2 3 4 5 7 7 10 0

    //简单向右旋转
    std::rotate(v.rbegin(), v.rbegin() + 1, v.rend());
    print("simple rotate right:\t", v);       // 0 1 2 2 3 4 5 7 7 10
}
```

6. std::unique

std::unique 可从[first, last)内的每组连续相同的元素中删除第一个元素之外的所有元素,并返回一个新的末尾迭代器。std::unique 有两个变体:

```
template<class ForwardIt>
ForwardIt unique( ForwardIt first, ForwardIt last);

template<class ForwardIt, class BinaryPred>
ForwardIt unique(ForwardIt first,ForwardIt last,BinaryPred p);
```

在第一种变体中,元素的比较操作使用的是"=="运算符。在第二种变体中,元素的比较操作使用的是二元谓词 p。

【例 6-13】

```cpp
#include <algorithm>
#include <iostream>
#include <vector>

int main(){
    std::vector<int> v{3, 6, 3, 3, 5, 5, 7, 4, 8, 8};
    auto print = [&](int id){
        std::cout << "@" << id << ": ";
        for (int i : v)
            std::cout << i << ' ';
        std::cout << '\n';
    };

    print(1);
    //删除 v 中相同元素
    auto last = std::unique(v.begin(), v.end());
    //v 现在变成{3  6  3  5  7  4  8  x  x  x},这里 x 是不确定的
    v.erase(last, v.end());                   //删除 v 中的 x
    print(2);
    //排序
    std::sort(v.begin(), v.end());            //v = {3  3  4  5  6  7  8}
    print(3);
    last = std::unique(v.begin(), v.end());
    //v 现在变成{3  4  5  6  7  8  x}
```

```
        v.erase(last, v.end());                //v = {3  4  5  6  7  8}
        print(4);
}
```

7. std::shuffle

std::shuffle 可对[first, last)内的元素进行重新排列，使这些元素的每个可能的排列都具有相等的出现概率。std::shuffle 的语法如下：

```
template< class RandomIt, class URBG >
void shuffle( RandomIt first, RandomIt last, URBG&& g);
```

这里 g 是生成器对象，返回一个随机选择的值。

【例 6-14】

```
#include <iostream>
#include <algorithm>
#include <array>
#include <random>
#include <chrono>

int main () {
    std::array<int,5> v{1, 2, 3, 4, 5};

    //生成一个时间种子
    unsigned seed = std::chrono::system_clock::now().time_since_epoch().count();
    shuffle(v.begin(),v.end(),std::default_random_engine(seed));

    std::cout << "shuffled elements:";
    for (int& x: v) std::cout << ' ' << x;
    std::cout << '\n';

    return 0;
}
```

6.1.5 分割算法

STL 算法库中的分割算法有 std::partition、std::is_partitioned、std::stable_partition、std::partition_copy、std::partition_point。限于篇幅，本节只介绍 std::partition。

std::partition 可重新排列[first, last)内的元素，使一元谓词 pred 返回 true 的所有元素都先于返回 false 的所有元素，并且保持元素间的相对顺序不变。迭代器返回指向第二组的第一个元素的位置。

```
template <class ForwardIterator, class UnaryPredicate>
ForwardIterator partition (ForwardIterator first, ForwardIterator last, UnaryPredicate pred);
```

【例 6-15】

```
#include <iostream>
#include <algorithm>
#include <vector>
```

```cpp
bool IsOdd (int i) { return (i%2)==1; }

int main (){
    std::vector<int> v{1,2,3,4,5,6,7,8,9};

    std::vector<int>::iterator bound;
    bound = std::partition(v.begin(), v.end(), IsOdd);

    std::cout << "奇数元素:";
    std::vector<int>::iterator it;

    for (it = v.begin(); it!=bound; ++it)
        std::cout <<' '<< *it;
    std::cout << '\n';

    std::cout << "偶数元素:";
    for (it=bound; it!=v.end(); ++it)
        std::cout <<' '<< *it;
    std::cout << '\n';
    return 0;
}
```

6.1.6 合并算法

STL 算法库中的合并算法有 std::merge、std::inplace_merge、std::includes、std::set_union、std::set_intersection、std::set_difference、std::set_symmetric_difference。限于篇幅，本节只介绍 std::merge。

std::merge 可将两个已经排序的[first1, last1)和[first2, last2)合并为一个从 d_first 开始的且已经排好序的范围。std::merge 有两个变体：

```cpp
template< class InputIt1, class InputIt2, class OutputIt >
OutputIt merge(InputIt1 first1, InputIt1 last1, InputIt2 first2, InputIt2 last2, OutputIt d_first );

template< class InputIt1, class InputIt2, class OutputIt, class Compare >
OutputIt merge( InputIt1 first1, InputIt1 last1, InputIt2 first2, InputIt2 last2, OutputIt d_first, Compare comp );
```

在第一个变体中，排序是按照 std::less 进行的，即升序排序。在第二个变体中，比较函数对象 comp 起排序作用。

std::merge 是稳定的，这意味着对于原来两个范围中的相等元素，第一个范围中的元素（保留其原始顺序）先于第二个范围中的元素（保留其原始顺序）。如果输出范围与[first1, last1)或[first2, last2)重叠，则程序运行结果将无法预测。请看下面的例子。

【例 6-16】

```cpp
#include <algorithm>
#include <functional>
#include <iostream>
#include <iterator>
#include <random>
```

```cpp
#include <vector>
auto print = [](const auto rem, const auto& v){
    std::cout << rem;
    std::copy(v.begin(), v.end(), std::ostream_iterator<int>(std::cout, " "));

    std::cout << endl;
};

int main(){
    //用随机数填充向量数组
    std::random_device rd;
    std::mt19937 mt(rd());
    std::uniform_int_distribution<> dis(0, 9);

    std::vector<int> v1(10), v2(10);
    std::generate(v1.begin(), v1.end(), std::bind(dis, std::ref(mt)));
    std::generate(v2.begin(), v2.end(), std::bind(dis, std::ref(mt)));

    print("Originally v1: ", v1);
    print("Originally v2: ", v2);

    std::sort(v1.begin(), v1.end());
    std::sort(v2.begin(), v2.end());

    print("After sorting v1: ", v1);
    print("After sorting v2: ", v2);

    //merge
    std::vector<int> dst;
    std::merge(v1.begin(), v1.end(), v2.begin(), v2.end(), std::back_inserter(dst));

    print("After merging dst: ", dst);
    return 0;
}
```

6.1.7 堆算法

使用 STL 可以在一定范围内实现堆数据结构，STL 提供了更快速的最大或最小项目检索，以及对排序后数据的更快速插入和删除，并且还可以用作堆排序的子程序。STL 算法库中的堆算法主要有 std::push_heap、std::pop_heap、std::make_heap、std::sort_heap、std::is_heap、std::is_heap_until。

- std::make_heap：将给定范围转换为堆。
- std::push_heap：在堆的末尾插入元素后重新排列堆元素。
- std::pop_heap：把 max 元素移到堆的末尾，以便删除 max 元素。
- std::sort_heap：按升序对堆元素进行排序。
- std::is_heap：检查给定的范围是否为最大堆。
- std::is_heap_until：返回最大堆的最大子范围。

限于篇幅，本节只介绍 std::make_heap 和 std::sort_heap。

1. std::make_heap

std::make_heap 可在[first, last)内构造一个堆，在默认情况下，会生成最大堆，但可以使用自定义的比较器将最大堆改为最小堆。堆是一种基于树的特殊数据结构，它满足堆属性，即堆的根节点的键值必须是堆中所有键值中的最大值（或最小值）。std::make_heap 有两个变体：

```
template< class RandomIt >
void make_heap( RandomIt first, RandomIt last );

template< class RandomIt, class Compare >
void make_heap( RandomIt first, RandomIt last, Compare comp);
```

第一种变体是默认的构造，使用的比较运算是 std::less。第二种变体使用指定的比较函数 comp。

【例 6-17】

```
#include <iostream>
#include <algorithm>
#include <vector>

int main () {
    int a[] = {3, 2, 4, 1, 5, 9};
    std::vector<int> v(a, a+6);

    std::make_heap (v.begin(),v.end());              //堆: 9 5 4 1 2 3
    std::cout << "initial max heap : " << v.front() << endl;

    std::pop_heap (v.begin(),v.end());               //堆: 5 3 4 1 2 9
    v.pop_back();
    std::cout << "max heap after pop : " << v.front() << endl;

    v.push_back(6);
    std::push_heap(v.begin(),v.end());               //堆: 6 5 3 4 1 2
    std::cout << "max heap after push: " << v.front() << endl;

    std::sort_heap (v.begin(),v.end());              //堆: 1 2 3 4 5 6
    for (size_t i=0; i<v.size(); i++)
        std::cout <<' '<< v[i];

    std::cout << endl;
    return 0;
}
```

2. std::sort_heap

std::sort_heap 可将[first, last)转换为一个按升序排序后的范围，并且无须再维护堆属性。std::sort_heap 的语法如下：

```
template< class RandomIt >
void sort_heap(RandomIt first, RandomIt last);
```

【例 6-18】

```cpp
#include <algorithm>
#include <iostream>
#include <string_view>
#include <vector>

int main(){
    std::vector<int> v{6, 3, 2, 5, 2, 8};
    vector<int>::iterator iter;
    is_heap(v.begin(), v.end()) ? cout << "The container is a heap " : cout << "The container is not a heap";
    std::make_heap(v.begin(), v.end());
    std::sort_heap(v.begin(), v.end());
    auto it = is_heap_until(v.begin(), v.end());
    for(iter = v.begin(); iter != it; iter++)
    cout << *iter << " ";
    return 0;
}
```

6.1.8 最大最小值算法

STL 算法库中的最大最小值算法有 std::max、std::max_element、std::min、std::min_element、std::minmax、std::minmax_element、std::next_permutation、std::prev_permutation、std::lixicographically_compare 等。限于篇幅，本节只介绍 std::lixicographically_compare 和 std::next_permutation。

1. std::lixicographically_compare

std::lixicographically_compare 可检查第一个范围[first1, last1)内的元素在字典上是否小于第二个范围[first2, last2)内的元素，如果是则返回 true。std::lixicographically_compare 有两个变体：

```cpp
template< class InputIt1, class InputIt2 >
bool lexicographical_compare(InputIt1 first1, InputIt1 last1, InputIt2 first2, InputIt2 last2);

template<class InputIt1, class InputIt2, class Compare>
bool lexicographical_compare(InputIt1 first1, InputIt1 last1, InputIt2 first2, InputIt2 last2, Compare comp);
```

在第一种变体中，元素比较使用的运算符是<；在第二种变体中，元素比较使用的比较函数是 comp。

【例 6-19】

```cpp
#include <algorithm>
#include <iostream>
#include <random>
#include <vector>

void print(const std::vector<char>& v, auto suffix){
    for (char c : v)
    std::cout << c <<' ';
    std::cout << suffix;
}
```

```cpp
int main(){
    std::vector<char> v1{'a', 'b', 'c', 'd'};
    std::vector<char> v2{'a', 'b', 'c', 'd'};

    std::mt19937 g{std::random_device{}()};
    for (;!std::lexicographical_compare(v1.begin(), v1.end(), v2.begin(), v2.end());)
    {
        print(v1, ">= ");
        print(v2, '\n');

        std::shuffle(v1.begin(), v1.end(), g);
        std::shuffle(v2.begin(), v2.end(), g);
    }
    print(v1, "<    ");
    print(v2, '\n');
}
```

2. std::next_permutation

std::next_permutation 可将范围[first, last)内的元素置换到下一个排列中。如果存在这样的下一个排列，则返回 true；否则，将范围[first, last)内的元素转换为字典上的第一个排列（如 std::sort）并返回 false。std::next_permutation 有两个变体：

```cpp
template<class BidirIt>
bool next_permutation(BidirIt first, BidirIt last);

template<class BidirIt, class Compare>
bool next_permutation(BidirIt first, BidirIt last, Compare comp);
```

在第一种变体中，集合是通过运算符"<"按字典顺序排列的。在第二种变体中，集合是通过比较函数 comp 按字典顺序排列的。

【例 6-20】

```cpp
#include <algorithm>
#include <iostream>
#include <string>

int main(){
    std::vector<std::string> s = {"cat", "dog", "cow"};
    do{
        std::cout << s[0] << s[1] << s[2] << '\n';
    }while (std::next_permutation(s.begin(), s.end()));

    std::cout << s[0] << s[1] << s[2] << '\n';
    std::sort(s.begin(), s.end());
    std::reverse(s.begin(), s.end());

    do{
        std::cout << s[0] << s[1] << s[2] << '\n';
    }while (std::prev_permutation(s.begin(), s.end()));
```

```
        return 0;
}
```

6.1.9 数值算法

STL 算法库中的数值算法有 std::iota、std::accumulate、std::partial_sum，本节介绍它们的应用。在使用这些算法时，需要在程序中添加"#include <numeric>"。

1．std::iota

std::iota 可以用顺序递增的值填充范围[first, last)，从 value 开始，重复执行++value 操作。std::iota 的语法如下：

```
template< class ForwardIt, class T >
void iota( ForwardIt first, ForwardIt last, T value );
```

【例 6-21】

```
#include <iostream>
#include <numeric>

int main(){
    std::array<int, 10> numbers;
    std::iota (numbers.begin(), numbers.end(),100);
    std::cout << "numbers: ";
    for (auto i= numbers.begin(); i < numbers.end(); ++i)
        std::cout << *i << ' ';
    std::cout << endl;
    return 0;
}
```

2．std::accumulate

std::accumulate 可计算[first, last)内的元素总和。std::accumulate 有两个变体：

```
template< class InputIt, class T >
T accumulate( InputIt first, InputIt last, T init );

template< class InputIt, class T, class BinaryOp >
T accumulate(InputIt first,InputIt last, T init,BinaryOp op);
```

第一种变体是基本格式，先使用 init 初始化[first, last)内的元素，再进行累加。第二种变体同样先使用 init 初始化[first, last)内的元素，再使用函数对象 op 进行累加。

【例 6-22】

```
#include <functional>
#include <iostream>
#include <numeric>
#include <string>
#include <vector>

int main(){
    std::vector<int> v{1, 2, 3, 4, 5, 6, 7, 8, 9, 10};
```

```cpp
    int sum = std::accumulate(v.begin(), v.end(), 0);
    int product = std::accumulate(v.begin(), v.end(), 1,
    std::multiplies<int>());

    auto dash_fold = [](std::string a, int b)
    {
        return std::move(a) + '-' + std::to_string(b);
    };

    std::string s = std::accumulate(std::next(v.begin()), v.end(), std::to_string(v[0]), dash_fold);
    std::string rs=std::accumulate(std::next(v.rbegin()), v.rend(), std::to_string(v.back()), dash_fold);
    std::cout << "sum: " << sum << '\n'
              << "product: " << product << '\n'
              << "dash-separated string: " << s << '\n'
              << "dash-separated string (right-folded): " << rs
              << '\n';
    return 0;
}
```

3. std::partial_sum

std::partial_sum 可以对[first, last)内的元素进行计算,然后把计算结果存放在输出序列以 d_first 为起始位置处。std::partial_sum 有两个变体:

```
template< class InputIt, class OutputIt >
OutputIt partial_sum(InputIt first, InputIt last, OutputIt d_first);
```

```
template<class InputIt, class OutputIt, class BinaryOp>
OutputIt partial_sum(InputIt first, InputIt last, OutputIt, d_first, BinaryOp op);
```

在第一种变体中,std::partial_sum 按顺序执行以下操作:

(1) 创建一个累加器 acc,其类型是 InputIt 类型,并使用*first 对其进行初始化。

(2) 将 acc 赋值给输出序列的*d_first。

(3) 对于[1, std::distance(first, last))中的每个整数 i,按顺序执行以下操作:

① 计算 std::move(acc) + *iter,iter 是迭代器。

② 将结果赋值给 acc。

③ 将 acc 赋值给*dest,其中 dest 是第 i 个 d_first 的迭代器。

第二种变体和第一种变体基本相同,只是使用 op(std::move(acc), *iter)进行计算,而不是 std::distance。

【例 6-23】

```cpp
#include <iostream>
#include <functional>
#include <numeric>

int add (int x, int y) {return x+y;}

int main () {
```

```cpp
        int v[] = {3,2,5,8,7,6};
        int result[6];

        std::partial_sum (v, v+6, result);
        std::cout << "using default partial_sum: ";
        for (int i=0; i<6; i++)
        std::cout << result[i] << ' ';
        std::cout << endl;

        //result 的值现在为:   3   5   10   18   25   31
        std::partial_sum (v, v+6, result, std::multiplies<int>());
        std::cout << "using functional operation multiplies: ";
        for (int i=0; i<6; i++)
        std::cout << result[i] << ' ';
        std::cout << endl;

        std::partial_sum (v, v+6, result, add);
        std::cout << "using custom function: ";
        for (int i=0; i<6; i++)
        std::cout << result[i] << ' ';
        std::cout << endl;

        return 0;
}
```

6.2 函数对象

函数对象（function object，又称仿函数）是 C++中通过重载函数调用运算符 operator()实现的可调用对象。例如：

```cpp
class operation {
    int operator() (int a, int b) {return a + b;}
};
operation myoperation;
int x = myoperation(2,3);
```

函数对象通常被当成参数传递给另一个函数，可作为 STL 中算法函数的谓词（predicate）或者比较函数（comparison function）。

头文件<functional>定义了 STL 中多个用于表示函数对象的类模板，包括算法操作、比较操作、逻辑操作，以及用于绑定函数对象实参的绑定器（binder）。这些类模板是具有函数调用运算符 operator()的 C++类，这些类的实例可以如同函数一样被调用。

头文件<functional>中的函数和函数对象主要分为：
- 函数模板：std::bind、std::cref、std::mem_fn、std::not1、std::not2、std::ref。
- 包装器类：std::function、std::binary_negate、std::reference_wrapper、std::unary_negate。
- 运算符类：std::bit_and、std::bit_or、std::bit_xor、std::divides、std::equal_to、std::greater、std::greater_equal、std::less、std::less_equal、std::logical_and、std::logical_not、std::logical_or、

std::minus、std::modulus、std::multiplies、std::negate、std::not_equal_to、std::plus。
- 其他类：std::bad_function_call、std::hash、std::is_bind_expression、std::is_placeholder。

运算符类中的都是函数对象，也就是说，它们都重载了函数调用运算符 operator()，在使用时可以直接调用重载函数。std::bind 是函数模板，std::function 是类包装器，它们在 6.4 节中详细介绍。限于篇幅，本节只介绍 std::greater、std::less、std::reference_wrapper、std::ref 和 std::cref。

6.2.1　std::greater 和 std::less

std::greater 和 std::less 是用于执行比较操作的函数对象，主函数只需在类型 T 上调用运算符>即可调用两个函数对象。std::greater 和 std::less 的语法如下：

```
template<class T = void>
struct greater;

template<class T = void>
struct less;
```

【例 6-24】

```cpp
#include <iostream>
#include <functional>
#include <algorithm>
#include <vector>

int main (){
    vector <int> vec {20, 40, 50, 10, 30};
    std::sort(vec.begin(), vec.end(), std::greater<int>());
    for (int i=0; i<5; i++)
    std::cout << vec[i] <<' ';
    std::cout << endl;

    vec.push_back(60);
    vec.push_back(35);
    std::sort(vec.begin(), vec.end(), std::less<int>());
    for (auto it = vec.begin(); it != vec.end(); ++it)
    std::cout << *it <<' ';

    std::cout << endl;
    return 0;
}
```

6.2.2　std::reference_wrapper

std::reference_wrapper 是集中类模板，用于包装对实例对象的引用，既可复制构造，又可拷贝赋值，其语法如下：

```
template <class T>
class reference_wrapper;
```

【例 6-25】

```cpp
#include <iostream>
#include <functional>
int main(){
    int a(10),b(20),c(30);
    //an array of "references":
    std::reference_wrapper<int> refs[] = {a,b,c};
    std::cout << "refs:";
    for (int& x : refs) std::cout <<' '<< x;
    std::cout << endl;
    return 0;
}
```

6.2.3　std::ref 和 std::cref

std::ref 用于构造一个适当的 std::reference_wrapper 类型的对象来保存对实例对象的引用。std::cref 用于构造一个适当的 std::reference_wrapper 类型的对象来保存对实例对象的 const 引用。std::ref 和 std::cref 的语法相同，下面给出的是 std::cref 的语法。

```cpp
template <class T>
reference_wrapper<const T> cref(const T& elem) noexcept;

template <class T>
reference_wrapper<const T> cref(reference_wrapper<T>& x)
noexcept;

template <class T>
void cref (const T&&) = delete;
```

【例 6-26】

```cpp
#include <iostream>
#include <functional>

void inc(int& x) {
    x++;
}

int main(){
    int foo (10);
    int bar (20);
    auto a = std::ref(foo);
    auto b = std::cref(bar);
    ++foo;
    ++bar;
    std::cout << a << endl;          //输出 11
    std::cout << b << endl;          //输出 21
```

```
    auto func1 = std::bind(inc, foo);
    func1();
    std::cout << foo << endl;              //输出 11
    auto func2 = std::bind(inc, std::ref(bar));
    func2();
    std::cout << bar << endl;              //输出 22
    return 0;
}
```

6.3 Utility 函数

STL 的<utility>库中有一组很有用途的功能性函数：std::move、std::forward、std::swap、std::make_pair、std::make_if_noexcept、std::declval。限于篇幅，本节只介绍前 4 个函数。

6.3.1 std::move

std::move 可在当前容器中移动元素并返回其右值引用，其语法如下：

```
template< class T >
typename std::remove_reference<T>::type&& move(T&& t) noexcept;

template< class T >
constexpr std::remove_reference_t<T>&& move(T&& t) noexcept;
```

【例 6-27】

```
#include <iostream>
#include <utility>
#include <vector>

int main () {
    std::string a = "nicky";
    std::string b = "Vicky";

    std::vector<std::string> name;

    //把 nicky 放入 vector 中，a 还是 nicky
    name.push_back(a);
    //把 Vicky 放入 vector 中，b 变成 NULL
    name.push_back(std::move(b));
    return 0;
}
```

类的移动构造和移动赋值重载就是使用 std::move 实现的，例如：

```
A(A&& arg) : member(std::move(arg.member)){}
A& operator=(A&& other){
    member = std::move(other.member);
```

```cpp
        return *this;
    }
```

6.3.2 std::forward

std::forward 需要使用模板类型 T，如果函数的参数 t 是左值引用的，则返回左值；如果不是左值引用的，则返回对 t 的右值引用。std::forward 用于保留传递给另一个函数的参数的正确值类别，该函数可以重载其参数的值类别，常用于包装器样式的委派。std::forward 的语法如下：

```cpp
template<class T>
T&&forward(typename std::remove_reference<T>::type& t)noexcept;

template<class T>
constexpr T&&forward(std::remove_reference_t<T>&& t)noexcept;
```

【例 6-28】

```cpp
#include <utility>
#include <iostream>

void g(const int& x) {
    std::cout << "[lvalue]";
}

void g (int&& x) {
    std::cout << "[rvalue]";
}

void func (int& x){
    g(std::forward<int &>(x));          //左值引用
}

void func (int&& x){
    g(std::forward<int &&>(x));         //右值引用
}

int main(){
    int a;
    std::cout << "左值引用 func：";
    func (a);
    std::cout << endl;

    std::cout << "右值引用 func: ";
    func (0);
    std::cout << endl;

    return 0;
}
```

6.3.3 std::swap

std::swap 用于交换两个给定变量的值或者两个给定数组的值，其语法如下：

```
template< class T >
void swap( T& a, T& b )noexcept;

template< class T2, std::size_t N >
void swap( T2 (&a)[N], T2 (&b)[N])noexcept;
```

【例 6-29】

```
#include <iostream>
#include <utility>

int main(){
    int x=10, y=20;
    std::swap(x,y);                    //x = 20、y=10
    int a[4];                          //a: ? ? ? ?
    int b[] = {10,20,30,40};           //b: 10 20 30 40
    std::swap(a,b);                    //a: 10 20 30 40; b: ? ? ? ?
    std::cout << "a contains: ";
    for (int i: a) std::cout << i << ' ';
    std::cout << endl;
    return 0;
}
```

6.3.4 std::make_pair

std::make_pair 用于创建一个 std::pair 对象，并从参数类型中推导出目标类型。std::make_pair 有两个变体：

```
template< class T1, class T2 >
std::pair<T1, T2> make_pair( T1 t, T2 u);

template< class T1, class T2 >
std::pair<V1, V2> make_pair( T1&& t, T2&& u);
```

在第二种变体中，V1 和 V2 是推导的类型，即 V1 和 V2 是 std::decay<T1>::type 和 std::decay<T2>::type。

【例 6-30】

```
#include <utility>
#include <iostream>

int main () {
    int n = 1;
    int arr[5] = {1, 2, 3, 4, 5};
```

```cpp
    std::pair<int,int> a;
    std::pair<int,int> b;

    a = std::make_pair(10,20);
    b = std::make_pair(10, 'A');              //正确：隐式转换

    auto p2 = std::make_pair(std::ref(n), arr);
    std::cout << "a: " << a.first << ", "<< a.second<< '\n';
    std::cout << "b: " << b.first << ", "<< b.second << '\n';

    n = 7;
    std::cout << "The value of p2 is " << '(' << p2.first << ", "<<*(p2.second+2)<<")\n";
    return 0;
}
```

6.4 回调函数

6.4.1 回调函数的基本概念

回调函数（callback function）是一种特殊的函数，是作为参数传递给被调用函数的函数。回调函数通常用于事件处理、异步编程，以及各种操作系统和框架的 API。

回调函数可以使用不同的语言工具实现。在 C++中，所有支持函数调用语义的实体都可称为可调用对象（callable object）。可调用对象可以是普通函数、函数指针、函数对象、lambda 表达式、std::bind、std::function。

本节将介绍使用上述可调用对象实现回调函数的方法。

6.4.2 使用普通函数实现回调函数

我们先看看直接用普通函数实现回调函数的简单例子。

【例 6-31】

```cpp
#include <iostream>
//回调函数
int ascii_code(char c){
    return (int)c;
}
void print(char c, int(*func_ptr)(char)) {
    int ascii = func_ptr(c);
    std::cout << "ASCII code of " << c << " is: " << ascii << endl;
}
int main()
{
    print('g', &ascii_code);
    return 0;
}
```

6.4.3 使用函数指针实现回调函数

在 C++语言中,通过函数指针可以将回调函数传递给其他函数。函数指针是一个变量,它存储了一个函数的地址。将函数指针作为参数传递给另一个函数时,这个函数就可以使用这个指针来调用其指向的函数。函数指针的形式如下:

返回类型 (*函数指针名称)(参数列表)

我们来看一个简单的回调函数,它有两个浮点形参并返回参数的最大值。函数指针的定义如下:

typedef float (*CALLBACK)(float, float);

这里的 CALLBACK 是函数指针类型,我们可以定义:

CALLBACK callback;

callback 是一个函数指针,可以让该函数指针指向特定的函数,如 "callback = &max;"。函数 max 的定义如下:

```
float max(float a, float b){
    return a> b? a : b;
}
```

接下来我们定义被调用函数。如上所述,这个被调用函数一般是事件驱动的或者消息触发的函数。这里我们只是用一个简单的函数,定义如下:

【例 6-32】

```
#include <iostream>
float do_math(float a,float b, float(*Callback)(float a,float b))
{
    return Callback(a,b);
}

int main()
{
    CALLBACK callback;
    callback = &max;
    float  a = do_math(5.5, 4.5, callback);
    cout << a;
    return 0;
}
```

6.4.4 使用函数对象实现回调函数

函数对象是一个类的实例,其中重载了函数调用运算符 operator()。当将一个函数对象作为参数传递给另一个函数时,另一个函数就可以使用这个对象来调用重载了的函数调用运算符 operator()。

假设有一个回调函数需要接收两个整型参数并返回一个整数值，可以使用以下方式定义函数对象。

【例 6-33】

```cpp
#include <iostream>
class MyFunctor
{
public:
    int operator( ) (int a , int b)                //重载函数调用运算符 operator()
    {
        return a+b;
    }
};

int do_math(void* context,int a,int b, int(*operation)(void* context,int a,int b)){
    return operation(context,a,b);
}

int main()
{
    MyFunctor functor;
    int a = do_math(&functor, 4, 6, MyFunctor);
    cout << a;
    return 0;
}
```

再看一个比较完整的例子。

【例 6-34】

```cpp
#include <iostream>
class Encryptor {
private:
    bool m_isIncremental;
    int m_count;
public:
    Encryptor() {
        m_isIncremental = 0;
        m_count = 1;
    }

    Encryptor(bool isInc, int count) {
        m_isIncremental = isInc;
        m_count = count;
    }

    //重载函数调用运算符 operator()
    std::string operator( ) (std::string data) {
        for (int i = 0; i < data.size(); i++)
            if ((data[i] >= 'a' && data[i] <= 'z') || (data[i] >= 'A' && data[i] <= 'Z'))
```

```cpp
            if (m_isIncremental)
                data[i] = data[i] + m_count;
            else
                data[i] = data[i] - m_count;
            return data;
        }
};
std::string doMessage(std::string rawData, Encryptor FuncObj){
    rawData = "[HEADER]" + rawData + "[FOOTER]";
    rawData = FuncObj(rawData);
    return rawData;
}

int main(){
    std::string msg = doMessage("ModernC++",Encryptor(true,1));
    std::cout << msg << std::endl;
    msg = doMessage("ModernC++", Encryptor(true, 2));
    std::cout << msg << std::endl;
    msg= doMessage("ModernC++", Encryptor(false, 1));
    std::cout << msg << std::endl;
    return 0;
}
```

6.4.5 将 lambda 表达式传入回调函数

在调用回调函数时，可以将 lambda 表达式传入回调函数。在 C++11 之前，如果要把一个函数作为实参传入其他函数，则需要使用函数指针或者函数对象。但是这样做比较烦琐，需要显式地定义函数或者类，代码的可读性不高。lambda 表达式是一个可调用对象，使用 lambda 表达式可以简化这个过程，使得代码更加简洁和易读。

【例 6-35】

```cpp
#include<iostream>
#include<vector>
#include<cstdlib>
//函数 Calculate()的参数列表里有一个 std::vector 和两个函数指针
void Calculate (const std::vector<int>& vec, const void (*calculate_square)(int),
                                             const void (*calculate_cube)(int)){
    for (const auto& num : vec) {
        if (num % 2 == 0) {
            calculate_square (num);
        } else {
            calculate_cube (num);
        }
    }
}

int main(){
```

```cpp
        srand(time(NULL));                    //种子随机函数
        std :: vector<int> vec_num;
        for (int i=0; i<10; i++) {
            vec_num.push_back(rand() % 20);
        }
        auto lambda_square =[](int num)->void{ std::cout<< "Number : " << num << " Square : "
                                                << num * num << std::endl; };
        auto lambda_cube =[](int num)->void{std::cout << "Number : " << num << " Cube : "
                                                << num * num * num << std :: endl; };
        Calculate (vec_num, &lambda_square, &lambda_cube);
        return 0;
}
```

6.4.6　使用 std::bind 实现回调函数

std::bind 可以通过绑定一个普通函数来创建新的函数对象，原普通函数中的参数使用占位符表示。std::bind 是函数模板，其语法如下：

```
template< class F, class... Args >
bind( F&& f, Args&&... args );

template< class F, class... Args >
constexpr bind( F&& f, Args&&... args );
```

这里 f 是可调用对象，args 是可调用对象的参数。std::bind 返回一个未指定类型的函数对象。下面通过 std::bind 产生函数对象，并实现回调函数。

【例 6-36】

```cpp
#include <iostream>
#include <functional>
#include <algorithm>

//为了使用占位符_1、_2…
using namespace std::placeholders;
int add(int first, int second){
    return first + second;
}

bool divisible(int num, int den){
    return num % den == 0;
}

int main(){
    auto first_add_func = std::bind(&add, _1, _2);
    std::cout << first_add_func(12, 5) << std::endl;

    auto second_add_func = std::bind(&add, 12, _2);
    std::cout << second_add_func(5) << std::endl;
```

```cpp
    int arr[10] = {1, 20, 13, 4, 5, 6, 10, 28, 19, 15};
    auto divisible_by_5 = std::bind(&divisible, _1, 5);

    int count = 0;
    for(int i=0; i<10; i++)
    if(divisible_by_5(arr[i]))
    count++;

    std::cout << count << std::endl;
    return 0;
}
```

下面的示例中，std::bind 产生新的函数对象可以用 std::function 表示。

【例 6-37】

```cpp
#include <functional>
#include <iostream>

void f(int& n1, int& n2, const int& n3){
    std::cout<<"In function: "<< n1<<' '<<n2<<' '<< n3<<'\n';
    ++n1;
    ++n2;
}

int main(){
    int n1 = 1, n2 = 2, n3 = 3;
    std::function<void()> bound_f = std::bind(f, n1, std::ref(n2),
    std::cref(n3));

    n1 = 10;
    n2 = 11;
    n3 = 12;

    std::cout<<"Before function: "<<n1<<' '<< n2 <<' '<< n3 <<'\n';
    bound_f();
    std::cout <<"After function: "<<n1<<' '<<n2<<' '<< n3 << '\n';
    return 0;
}
```

6.4.7 使用 std::function 实现回调函数

C++11 引入了 std::function，它是一个通用多态函数封装器。std::function 的实例可以存储、复制和调用任何拷贝构造（copy constructible）的可调用目标（callable target），如函数指针、lambda 表达式、std::bind 或其他函数对象，以及指向成员函数的指针和指向数据成员的指针。使用 std::function 实现回调函数比使用函数指针或指向成员函数的指针更通用，因为使用 std::function 可以传递不同类型的对象并将它们隐式转换为 std::function 对象。std::function 的定义如下：

```cpp
template< class R, class... Args >
class function<R(Args...)>;
```

这里的 R 是函数返回值的数据类型，Args 是函数参数列表。使用 std::function 时需要加上 "#include <functional>"。

现在我们可以定义：

```cpp
typedef    class function<int (int)>    CALLBACK;
using CALLBACK = class function<int (int)>;
```

然后定义：

```cpp
CALLBACK callback;
```

这样就可以把任何返回值类型为 int 且只有一个整型参数的函数赋值给 callback，从而将该函数作为回调函数。

【例 6-38】

```cpp
#include <functional>
#include <iostream>

typedef std::function<int(int，int)> CALLBACK;
class CallbackExample{
public:
    void setCallback(CALLBACK cb){
        m_cb = cb;
    }

    void test() {
        std::cout << "CallbackExample::test() calling callback." << endl;
        int a = m_cb(10, 12);
        std::cout << "10 + 12 = " << a << endl;
    }

private:
    CALLBACK m_cb;
};
int Add(int x, int y) {return    x+y; }
int main(){
    CALLBACK cb = Add;
    CallbackExample cbexample;
    Cbexample.setCallback(cb);
    Cbexample.test();
    return 0;
}
```

下面的例子使用 std::function 函数模板作为被调用函数的参数。相较于函数指针而言，使用 std::function 实现回调函数更具通用性。

【例 6-39】

```cpp
#include <functional>
#include <iostream>

template<class R, class T>
```

```cpp
void mytransform(int* v, int n, std::function<R(T)> fp){
    for (int i = 0; i < n; ++i){
        v[i] = fp(v[i]);
    }
}

int square (int x) { return x*x;}
int cube (int x) { return x*x*x;}
char encode (char ch) {return ch + 8;}

typedef std::function<int(int)> CALLBACK;
int main(){
    int a[6] = {1, 2, 3, 4, 5, 6};
    float b[6] = {1.0, 2.0, 3.0, 4.0, 5.0, 6.0};
    char c[6] ={'A', 'B', 'C', 'D', 'E', 'F'};

    CALLBACK callback = square;
    mytransform<int,int>(&a[0],6,&callback);

    callback = cube;
    mytransform<float,float>(&b[0],6,&callback);
    mytransform<char,char>(&c[0],6,&encode);
    return 0;
}
```

最后一个例子是使用 std::function 作为回调函数，用于设计模式中的观察者（Observer）模式。对观察者设计模式不熟悉的读者可以先阅读本书第 10 章的相关内容。

```cpp
class Subject{
public:
    void AttachObserver(Observer* observer){
        observers.push_back(observer);
    }
    //...
};

using ObserverID = std::size_t;
std::mutex Mutex;

template<typename T>
class Observer{
public:
    using CallbackFunction = std::function<void(T)>;

public:
    ObserverID AttachObserver(CallbackFunction callback){
        std::unique_lock<std::mutex> block(Mutex);
        ObserverID id = GetNewID();
        _observers.emplace(id, std::move(callback));
```

```cpp
            return(id);
        }

        bool DetachObserver(ObserverID& id){
            std::unique_lock<std::mutex> block(Mutex);
            ObserverID it = _observers.find(id);
            if (it == _observers.end()){
                return false;
            }
            Observers_.erase(it);
            return(true);
        }

    protected:
        void NotifyAllObservers(T arg){
            std::unique_lock<std::mutex> block(Mutex);
            for (const auto& observer : observers_){
                observer_.second(arg);
            }
        }
    private:
        ObserverID GetNewID(){
            typedef void (*func_type) (T);
            static std::hash<func_type> hash;
            const ObserverID id = hash(callback);
            return id;
        }
        std::unordered_map<ObserverID,CallbackFunction>observers_;
};

class IntPublisher : public Observer<int> {
public:
    void Publish(int number) {
        NotifyAllObservers(number);
    }
};

IntPublisher publisher;
ObserverID id = publisher.AttachObserver([](int number)->void{ std::cout << number << std::endl;});
publisher.Publish(42);
publisher.DetachObserver(id);
```

6.5 本章小结

本章主要介绍 STL 的几个重要库（如<algorithm>、<functional>、<numeric>和<utility>）的算法函数和实用功能的函数，因此在使用相应的算法函数时，应加上相应的"#include <库名称>"。

在每一个库中，本章仅介绍了一些重要的算法函数，并给出了示例代码。限于篇幅，本章并没有逐一介绍所有的算法函数。

在<algorithm>中，每一个算法函数都是模板函数，在迭代器的帮助下，可以应用于各类容器，因此算法函数的通用性好。很多算法函数的最后一个参数是一元谓词或二元谓词，它们本质上都是一个可调用的函数对象。

函数对象通常是指重载了函数调用运算符的类，使得类可以像函数一样使用。<functional>中定义了一些有关逻辑操作、关系比较的运算符类（如 std::greater、std::less）等。函数对象通常作为参数传递给一些算法函数，如 std::sort。

回调函数是实际编程中经常使用的一种程序设计方法，通常应用于消息、通信、事件驱动等场合。本章详细介绍了回调函数的各种实现方法，并给出了每种实现方法的示例代码。

C++17 引入了并行计算的概念，<algorithm>中的一些算法也具有并行执行策略。通常，引用并行执行策略算法函数的第一个参数是 ExecutionPolicy。具有并行执行策略的算法函数将在第 8 章中介绍。

第 7 章 智能指针与内存管理

7.1 堆栈和内存分配

在 C++程序中,所有的变量都需要在内存中分配存储空间,如图 7.1-1 所示。程序中一般有常量、全局变量、静态变量、局部变量、函数参数等。栈(stack)和堆(heap)是两种主要的内存资源。栈是地址连续的内存块,是一种暂时性的内存资源。所有的函数参数都是在栈中分配存储空间的,所有函数内部定义的临时变量也是在栈中分配存储空间的。当函数运行结束时,函数参数和函数内部的临时变量在栈中分配的存储空间都将被系统回收。堆是由程序员通过特定的指令创建对象时为所创建对象分配的存储空间的总称。与栈不同,在堆中创建的对象不会被自动销毁,而且创建的对象对所有的线程都是可见的。如果没有销毁在堆中创建的对象,就会造成内存泄漏。严重的内存泄漏会导致程序运行变慢甚至崩溃,是 C++软件开发过程中遇到的最严重问题之一。

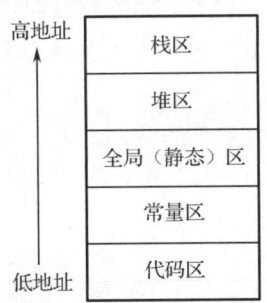

图 7.1-1　C++程序中变量在内存中分配的存储空间

我们通过一个小例子熟悉一下两个概念:栈溢出与堆溢出。

【例 7-1】
```
#include<stdio.h>
foo(int val, int i)
{
    int a[100000][100000];          //导致栈溢出
    //...
}

int main(){
    for (long int i=0; i<1000000000; i++)
    {
        int *ptr = new int();       //导致堆溢出
        *ptr = 10;
```

```
            foo(*ptr, i);
        }
    return 0;
}
```

在 C++中，我们使用 new 和 delete 两个指令在堆中创建对象和销毁对象。当我们使用 new 来创建对象时，就需要用一个变量来引用我们创建的对象，这个变量就是指针。这里的指针是被创建对象在堆中地址的符号表达形式。指针本身是在栈中存放的，但它指向的是在堆中创建的对象。例如：

```
string* pstr = new string("hello world.");
```

上面的语句在堆中创建了一个 string 对象，并且 string 被初始化为 hello world。指针 pstr 指向 string 对象在堆中的起始地址，但 pstr 本身存放在栈中，只是 pstr 的存储单元里的内容是 string 对象的地址。使用 delete pstr 可删除 pstr 指向的对象 string，pstr 本身在栈中，会被自动回收。

7.2 指针与内存泄漏

在 C++中使用指针时需要特别小心，不当使用指针可能带来如下的问题：
- 内存泄漏（memory leak）：重复分配内存而没有释放内存，会导致过多的内存消耗，最终导致系统崩溃。
- 悬挂指针（dangling pointer）：是指指针在释放内存后，没有赋给新的地址，没有置空。
- 野指针（wild pointer）：是指指针变量被声明后，从来没有被初始化指向任何对象或者地址。
- 数据不一致性（data inconsistency）：数据不一致性发生在内存中的一些数据没有用一致性方法进行更新。
- 缓存溢出（buffer overflow）：当使用指针将数据写入缓存时，指针越界内存块会导致数据崩溃，并且有可能被恶意利用。

下面列举几个初级程序员容易出现内存泄漏问题的示例。

【例 7-2】

```
#include <iostream>
using namespace std;
#define PI 3.1415926

class Circle{
public:
    Circle():x(0),y(0),radius(1.0){}
    Circle(x0,y0,r):x(x0),y(y0),radius(r){}
    double area()const {return PI*radius*radius;}
    double perimeter()const {return 2.0*PI*radius;}
private:
    double radius;
    double x;
    double y;
};

void generate_circle(double r){          //该函数中指针没有被删除，将造成内存泄漏
```

```cpp
        Circle* p = new Circle(5.0, 5.0, r);
        cout << "area=" << p->area() << endl;
}

int main(){
    int a=1;
    while(1){
        generate_circle(a);              //在循环内被调用,将造成严重的内存泄漏
        if(a>10)break;
        a++;
    }
    return 0;
}
```

下面的情况更为普遍,是不少初级程序员容易犯的错误。

```cpp
void MyFunction(){
    MyClass *obj = new MyClass(10);
    if(obj){
        //...
        int ret=obj->someOtherFunction();
        if(ret<0)
            return;                      //导致内存泄漏
    }
    //...
    delete obj;
}

double division(int numerator, int denominator){
    try{
        double res;
        if (denominator == 0) {
            throw runtime_error("Division by zero is not allowed!");
        }
        res = (double)numerator / denominator;
    }
    catch (const exception& e) {
        cout << "Exception " << e.what() << endl;
        res = -1.0;
    }
    return  res;
}

void MyMethod(){
    MyClass *obj = new MyClass;
    double ret = division();             //可能会抛出异常
    delete obj;
}
```

上面例子中的问题更隐蔽,如果 division()抛出异常,则指针对象 obj 无法被删除,将造成内存泄漏。下面我们再来看看面向对象编程中虚函数的应用可能带来的内存泄漏问题。

```cpp
class Base{
public:
    Base() { p = new int(5);}              //动态分配一个整型指针并初始化为 5
    ~Base() { delete p;}                    //在析构函数中删除指针
private:
    int * p;
};

class Derived: public Base{
public:
    Derived (){ pstr = new char[30];}      //动态分配 30 个字符的数组
    ~Derived (){ delete [ ] pstr;}          //删除分配 30 个字符的数组
private:
    char * pstr;
};

main(){
    Base * p = new Derived;
    //...
    delete p;
}
```

在上面的程序中，基类和派生类都在析构函数中删除所分配的指针，但由于基类的析构函数不是虚函数，因此在执行"delete p;"时，仅仅调用了基类的析构函数，没有传递到派生类，导致派生类中的指针变量未被删除，造成内存泄漏。

上面的几个例子介绍了指针带来的内存泄漏问题。内存泄漏是指针最为突出的问题，为此现代 C++ 引入了智能指针，并将智能指针分成几种不同的类型，保留了以前使用指针的习惯和传统，但无须使用 delete 来删除指针。

7.3 分段错误

在 C 或 C++ 程序中，经常会发生分段错误（segmentation fault），如图 7.3-1 所示。分段错误是指当程序尝试访问它无权访问的内存空间时所发生的错误。通常，当内存访问受到侵犯时，也会发生此错误，这是一种常规保护故障。

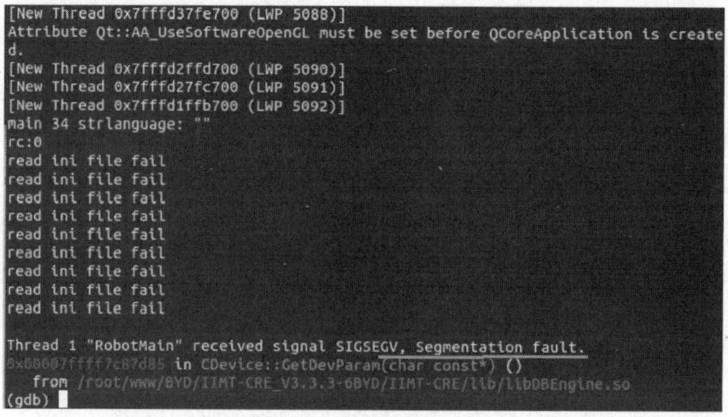

图 7.3-1 分段错误示例

程序在运行时可以访问特定的内存空间。栈用于保存每个函数的局部变量，程序可能在运行时分配内存并保存在堆上（在 C++程序中是 new 操作）。允许程序访问的唯一内存是它自己的内存空间，访问该内存空间之外的任何内存都会导致分段错误。发生分段错误通常是以下的几种错误导致的：

- 修改字符串文本。
- 访问已释放的地址。
- 访问数组的索引越界。
- scanf()使用不当。
- 堆栈溢出。
- 访问未初始化的指针。

分段错误的解决方案取决于发生错误的原因。要了解错误的原因，我们可以使用 gdb 调试器或 Valgrind 工具等。以下是由常见原因引发分段错误的专门解决方案。

- 初始化指针为 NULL，并在使用指针时检查指针是否为 NULL。
- 使用向量而不是数组来防止越界访问。
- 递归时避免堆栈溢出。
- 使用智能指针。

7.4 智能指针

我们先通过示例看看智能指针是如何设计的。

【例 7-3】

```cpp
#include <iostream>
using namespace std;

class SmartPointer {
    int* ptr;                    //实际指针
public:
    explicit SmartPointer(int* p = NULL) { ptr = p;}
    ~SmartPointer() { delete ptr; }

    int& operator*() { return *ptr; }
    int* operator->(){ return   ptr; }
}

int main(){
    SmartPointer p(new int(0));
    *p = 10;
    cout << *p;
    return 0;
}
```

上面的例子是智能指针的雏形。我们看到，智能指针是一个类，它把真正要创建的对象指针 int *ptr 封装起来，在构造函数中获取创建对象的地址并将该地址赋值给 ptr，在析构函数中删除

ptr。当我们使用智能指针时,感觉与传统指针没有差别,除了不再需要使用 delete。下面我们引进模板概念,对上面的示例进行扩展。

【例 7-4】

```cpp
#include <iostream>
using namespace std;

template <class T>
class SmartPointer {
public:
    explicit SmartPointer(T* p = NULL) { ptr = p; }
    ~SmartPointer() { delete ptr; }
    T& operator*() { return *ptr; }
    T* operator->() { return ptr; }
private:
    T* ptr;                          //实际指针
};

int main(){
    SmartPointer<int> p1(new int(0));
    *p1 = 20;
    cout << *p1;
    SmartPointer<double> p2(new double(10.0));
    cout << *p2;
    return 0;
}
```

自 C++11 起,STL 提供并实现了 4 种智能指针:std::auto_ptr、std::unique_ptr、std::shared_ptr、std::weak_ptr。std::auto_ptr 已被放弃,这里不再赘述。本节将详细介绍其他三种智能指针的特点、使用方法和注意事项。

7.4.1　std::unique_ptr

std::unique_ptr 只能存储一个对象,具有独占性。如果要放弃 std::unique_ptr 所指的对象,必须使用 std::move 将 std::unique_ptr 剥离所指的对象,其他指针才能指向该对象。在 C++11 中,std::unique_ptr 是一个模板类,定义在头文件<memory>中,因此需要在程序中添加 "#include <memory>"。std::unique_ptr 的声明如下:

```cpp
template< class T, class Deleter = std::default_delete<T>>
class unique_ptr;
```

其中:
class T:由用户定义的数据类型。
class Deleter:用户可以定义 Deleter,如果用户没有定义,则使用 std::default_delete<T>。
下面我们通过示例来展示 std::unique_ptr 的使用方法:

【例 7-5】

```cpp
#include <iostream>
#include <memory>
```

```cpp
#define PI 3.1415926
using namespace std;

class Circle {
    double x;
    double y;
    double radius;

public:
    Circle(double x0, double y0, double r0): x(x0),y(y0), radius(r0){ }
    double area() { return PI*r*r; }
    double perimeter() {return 2.0*PI*r;}
};

int main(){
    unique_ptr<Circle> p1(new Circle(5.0, 5.0, 10.0));
    cout << p1->area() << endl;
    cout << p1->perimeter()<<endl;

    unique_ptr<Circle> p2;
    p2 = move(p1);

    cout << p2->area() << endl;
    cout << p2->perimeter() << endl;
    return 0;
}
```

在例 7-5 中，p1 是一个 std::unique_ptr，在初始化后指向一个动态创建的对象 Circle(5.0,5.0,10.0)；p2 也是一个 std::unique_ptr，但我们不可以使用"p2 = p1;"，必须使用"p2 = move(p1);"，其中 move(p1) 剥离 p1 所指的对象，返回对象的地址，让 p2 指向对象 Circle(5.0,5.0,10.0)。这时，p1 没有指向任何对象了，是一个空指针。我们注意到，在 main() 的结尾并没有释放 p2，这是因为智能指针可通过自己的析构函数释放对象所分配的内存空间。

C++14 引入的 std::make_unique 可创建动态对象，其语法为：

std::make_unique <object_type> (arguments);

这里：
object_type：要创建的对象类型。
arguments：对象的构造函数参数列表。

【例 7-6】

```cpp
#include <iostream>
#include <memory>

struct Vec3{
    int x, y, z;
    Vec3() : x(0), y(0), z(0) {}
    Vec3(int x0, int y0, int z0) :x(x0), y(y0), z(z0) {}
    friend std::ostream& operator<<(std::ostream& os, Vec3& v){
```

```cpp
            return os << '{' << "x:" << v.x << " y:" << v.y << " z:" << v.z   << '}';
        }
};

int main(){
    std::unique_ptr<Vec3> v1 = std::make_unique<Vec3>();
    std::unique_ptr<Vec3> v2 = std::make_unique<Vec3>(0, 1, 2);
    std::unique_ptr<Vec3[]> v3 = std::make_unique<Vec3[]>(5);

    std::cout << "make_unique<Vec3>(): " << *v1 << '\n'
            << "make_unique<Vec3>(0,1,2): " << *v2 << '\n'
            << "make_unique<Vec3[]>(5): " << '\n';
    for (int i = 0; i < 5; i++) {
        std::cout << v3[i] << '\n';
    }
    return 0;
}
```

在例 7-6 中，v1、v2 和 v3 是三个 std::unique_ptr，分别存储三个动态创建的 Vec3 对象。当程序结束时，三个动态创建的 Vec3 对象内存空间都被释放掉。在本例中，我们再也没有看到 new 和 delete 了。

```cpp
void foo(std::unique_ptr<std::string> cp){
    //...
}
auto up = std::make_unique<std::string>("some strings");
foo(up)                                              //错误：不能使用拷贝方式
foo(std::move(up));                                  //正确
foo(std::make_unique<std::string>("some strings"));  //正确
```

我们需要牢记，std::unique_ptr 对存储对象的独占性，up 是一个 std::unique_ptr，并且被初始化后指向一个 string 对象，不可以再让 cp 共享，只有先使用 std::move 剥离 up，才可以让 cp 独占 string 对象。

在 std::unique_ptr 的定义中，我们提到过 Deleter 也是可以由用户自己定义的。下面看看用户如何定义自己的 Deleter。

【例 7-7】

```cpp
#include <iostream>
include <memory>

template <typename T>
class MyDeleter{
public:
    void operator()(T* ptr) const
    {
        std::cout << "freeing memory using 'delete'...\n";
        delete ptr;
    }
};
```

```cpp
template <typename T>
class MyDeleter<T[]>{
public:
    template <typename U>
    void operator()(U* ptr) const {
        std::cout << "freeing memory using 'delete[]'...\n";
        delete[] ptr;
    }
};
int main(){
    std::unique_str p1 = new int(0);
    std::unique_str p2 = new int[3]{5, 7, 8};
    std::unique_ptr<int, MyDeleter<int>> upi(std::move(p1), MyDeleter<int>{});
    std::unique_ptr<int[], MyDeleter<int[]>> upari(std::move(p2), MyDeleter<int[]>{});
    return 0;
}
```

上面的例子中使用 new 来创建数组对象，同样可以使用 std::make_unique 来创建数组对象，其语法格式为：

```
template< class T >
unique_ptr<T> make_unique( std::size_t size );
```

这里 T 为数组类型 U[]，数组的大小由 size 指定。

【例 7-8】

```cpp
#include <cstddef>
#include <iomanip>
#include <iostream>
#include <memory>
#include <utility>

struct Vec3{
    int x, y, z;
    Vec3(int x0=0,int y0=0,int z0=0)noexcept:x(x0),y(y0),z(z0){}
    friend std::ostream& operator<<(std::ostream& os,const Vec3& v)
    {
        return os<<"{x="<<v.x<<", y="<<v.y<<", z="<<v.z<<" }";
    }
};

int main(){
    //创建一个 std::unique_ptr 指向的包含 5 个元素的数组 Vec3[5]
    std::unique_ptr<Vec3[]> v3 = std::make_unique<Vec3[]>(5);
    for (std::size_t i = 0; i < 5; ++i)
        std::cout << std::setw(i? 30 : 0) << v3[i] << endl;
    std::cout << endl;
    return 0;
}
```

下面我们通过例子介绍 std::unique_ptr 的主要成员函数：reset、release 和 swap。

【例 7-9】

```cpp
#include <iostream>
#include <memory>
struct foo{
    foo() { std::cout << "foo...\n"; }
    ~foo() { std::cout << "~foo...\n"; }
};

struct D{                                        //自定义 Deleter
    void operator()(foo* p){
        std::cout << "Calling delete for foo object... \n";
        delete p;
    }
};

int main(){
    std::cout << "Creating a new foo \n";
    std::unique_ptr<foo, D> up(new foo(), D());

    std::cout << "Replace existing foo with a new foo \n";
    up.reset(new foo());             //调用 Deleter 删除已有的 foo

    std::cout << "Delete the owned foo and set to nullptr \n";
    up.reset(nullptr);               //调用 Deleter 删除 foo，并置空
    return 0;
}
```

在例 7-9 中，成员函数 reset 首先销毁智能指针所存储的对象，然后传入新创建的对象以取代现有的存储对象。也可以传入空指针，这样可起到纯粹删除所存储对象的作用。同时，例 7-9 采用了自定义的 Deleter。

【例 7-10】

```cpp
#include <cassert>
#include <iostream>
#include <memory>

struct foo{
    foo() { std::cout << "foo\n"; }
    ~foo() { std::cout << "~foo\n"; }
};

//转移 foo 的资源拥有权后将其销毁
void legacy_api(foo* owning_foo){
    std::cout << __func__ << '\n';      //遗留代码不能碰
    delete owning_foo;                  //销毁对象
}

int main(){
```

```cpp
    std::unique_ptr<Foo> managed_foo(new Foo);
    legacy_api(managed_foo.release());
    assert(managed_foo == nullptr);
    return 0;
}
```

成员函数 release 可释放对存储对象的拥有权，但不销毁对象。和 std::move 类似，release 返回的指针正是 std::unique_ptr 所指的对象。

【例 7-11】

```cpp
#include <iostream>
#include <memory>
struct foo{
    foo(int _val) : val(_val) { std::cout << "foo...\n"; }
    ~foo() { std::cout << "~foo...\n"; }
    int val;
};

int main(){
    std::unique_ptr<foo> s1(new foo(1));
    std::unique_ptr<foo> s2(new foo(2));
    s1.swap(s2);
    std::cout << "s1->val:" << s1->val << '\n';
    std::cout << "s2->val:" << s2->val << '\n';
    return 0;
}
```

成员函数 swap 的作用非常直观，可直接交换两个 std::unique_ptr 的存储对象。同时，我们看到 s1->val 和 s2->val，即使用指针的方式访问成员变量。这是因为 std::unique_ptr 的成员函数中定义了重载赋值运算符 operator* operator->，保留了指针的使用习惯。

【例 7-12】

```cpp
#include <iostream>
#include <memory>

int main(){
    std::unique_ptr<int> ptr(new int(42));
    if(ptr)
        std::cout << "before reset, ptr is: " << *ptr << endl;
    ptr.reset();    // 智能指针可以使用 "." 来访问成员变量和成员函
    ptr?(std::cout <<"after reset, ptr is: " << *ptr):(std::cout << "after reset ptr is empty") << endl;
    return 0;
}
```

例 7-12 应用了重载赋值运算符 operator bool，这样我们就可以使用指针的方式，如 if(ptr)和表达式中的 ptr?，以及*ptr。总之，智能指针既可以使用 "." 来访问成员变量和成员函数，也可以使用 "->" 来访问成员变量和成员函数，保留了指针的使用习惯。

从 C++20 起，C++引入了 std::make_unique_for_overwrite()，这里的后缀 "_for_overwrite" 用新创建的对象替代现有 std::unique_ptr 的内容。std::make_unique_for_overwrite()分为支持单个对象

和支持数组对象两种格式，支持单个对象的语法为：

```
template< class T >
unique_ptr<T> make_unique_for_overwrite();
```

这里，T 为非数组类型。支持数组对象的语法为：

```
template< class T >
unique_ptr<T> make_unique_for_overwrite(std::size_t size);
```

这里，T 为数组类型 U[]，数组大小由 std::size_t size 确定。

【例 7-13】

```cpp
#include <cstddef>
#include <iomanip>
#include <iostream>
#include <memory>
#include <utility>

struct Vec3{
    int x, y, z;
    Vec3(int x0=0,int y0=0,int z0=0) noexcept:x(x0),y(y0),z(z0){}
};

//输出斐波那契序列到迭代器
template<typename OutputIt>
OutputIt fibonacci(OutputIt first, OutputIt last){
    int a = 0;
    int b = 1;
    for (; first != last; ++first){
        *first = b;
        b += std::exchange(a, b);
    }
    return first;
}

int main(){
    std::unique_ptr<Vec3> v1 = std::make_unique<Vec3>();
    std::unique_ptr<Vec3> v2 = std::make_unique<Vec3>(5, 3, 2);
    std::unique_ptr<Vec3[]> v3 = std::make_unique<Vec3[]>(5);
    std::unique_ptr<int[]> v4 = std::make_unique_for_overwrite<int[]>(10);

    //生成斐波那契序列
    fibonacci(v4.get(), v4.get() + 10);
    std::cout << "v4: [" << v4[0];
    for (std::size_t i = 1; i < 10; ++i)
        std::cout << ", " << v4[i];
    std::cout << "]" << endl;

    return 0;
}
```

在例 7-13 中，std::make_unique_for_overwrite<int[]>(10)创建了一个包含 10 个整数的数组并且由智能指针 v4 管理。但 v4 不是 STL 的容器类，不能通过 v4.begin()、v4.end()将 v4 直接传递给 STL 的算法库函数，只能使用 v4.get()、v4.get()+10 传递给 STL 的算法库函数。

7.4.2　std::shared_ptr

使用 std::shared_ptr 可以让多个指针同时指向同一个对象，即共享方式。当对象被多个指针共享时，只有最后一个智能指针超出其作用范围后，动态分配的对象才会被销毁。在 C++11 中，std::shared_ptr 是一个类模板，定义在头文件<memory>中，因此需要在程序中添加"#include <memory>"。std::shared_ptr 的语法如下：

```cpp
template <class T>
class shared_ptr;
```

【例 7-14】

```cpp
#include <iostream>
#include <memory>
using namespace std;

class Circle {
    double x;
    double y;
    double radius;

public:
    Circle(double x0, double y0, double r): x(x0),y(y0),radius(r){}
    double area() { return PI * r * r; }
    double perimeter() {return 2.0 * PI * r;}
};

int main(){
    shared_ptr <Circle> p1( new Circle(5.0, 5.0,    10.0 ));
    cout << p1->area() << endl;
    cout << p1->perimeter() << endl;

    shared_ptr<Circle> p2;
    p2 = p1;

    cout << p2->area() << endl;
    cout << p2->perimeter() << endl;
    cout << p1->area() << endl;
    cout << p1->perimeter() << endl;
    cout << p1.use_count() << endl;            //输出 2，因为引用次数是 2
    return 0;
}
```

在例 7-14 中，p1 和 p2 是 std::shared_ptr，都同时指向对象 Circle(5.0, 5.0, 10.0)，它们都可以访问类 Circle 的 public 部分的函数 area()和 perimeter()，体现了共享特性。同样，在 main()的末尾

不需要手动删除 p1 和 p2，它们将随着 main()的结束自动调用自己的析构函数释放所分配的内存。

C++11 引入的 std::make_shared 可用来创建单个 std::shared_ptr 动态对象，但直到 C++20 才对 std::make_shared 进行了扩展，使 std::make_shared 可以指向数组对象，并同时引入了 std::make_share_for_overwrite。

表 7.4-1 展示了 std::make_shared 和 std::make_share_for_overwrite 的定义。

表 7.4-1　std::make_shared 和 std::make_share_for_overwrite 的定义

C++版本	定　　义
C++11	template< class T, class... Args > shared_ptr<T> make_shared(Args&&... args);
C++20 （T 的类型是 U[ ]）	template< class T > shared_ptr<T> make_shared(std::size_t N);
C++20 （T 的类型是 U[N]）	template< class T > shared_ptr<T> make_shared();
C++20 （T 的类型是 U[ ]）	template< class T > shared_ptr<T> make_shared(std::size_t N, const std::remove_extent_t<T>& u);
C++20 （T 的类型是 U[N]）	template< class T > shared_ptr<T> make_shared(const std::remove_extent_t<T>& u);
C++20 （T 的类型不是 U[ ]）	template< class T > shared_ptr<T> make_shared_for_overwrite();
C++20 （T 的类型是 U[ ]）	template< class T > shared_ptr<T> make_shared_for_overwrite(std::size_t N);

【例 7-15】

```
#include <iostream>
#include <memory>
#include <type_traits>
#include <vector>
using namespace std;

struct C{
    C(int i) : i(i) {}
    C(int i, float f) : i(i), f(f) {}
    int i;
    float f;
};

int main(){
    shared_ptr<C> sp1 = make_shared<C>(1);
    cout << "sp1->{ i:" << sp1->i << ", f:" << sp1->f << "} \n";

    shared_ptr<C> sp2 = make_shared<C>(2, 3.0f);
    cout << "sp2->{ i:" << sp2->i << ", f:" << sp2->f << "} \n";

    //std::shared_ptr 指向一个有初始值的浮点数数组 float[64]
    shared_ptr<float[]> sp3 = make_shared<float[]>(64);
    //std::shared_ptr 指向一个有初始值的 short[128]
```

```
    shared_ptr<short[128]> sp4 = make_shared<short[128]>();
    //std::shared_ptr 指向一个有初始值的 int[7][6][5]
    shared_ptr<int[7][6][5]> sp5 = make_shared<int[7][6][5]>();
    //std::shared_ptr 指向一个有初始值的 double[256]，每个元素的值都是 2.0
    shared_ptr<double[]> sp6 = make_shared<double[]>(256, 2.0);
    //std::shared_ptr 指向数组 double[7][2]，每一行元素都是{3.0, 4.0}
    shared_ptr<double[][2]> sp7 = make_shared<double[][2]>(7, {3.0, 4.0});
    //std::shared_ptr 指向向量数组 vector<int>[4]，每一个向量的初始值都是{5, 6}
    shared_ptr<vector<int>[]>sp8 = make_shared<vector<int>[]>(4,{5,6});
    //std::shared_ptr 指向数组 float[512]，每一个元素的值都是 1.0
    shared_ptr<float[512]> sp9 = make_shared<float[512]>(1.0);
    //std::shared_ptr 指向二维数组 double[6][2]，每一行元素都是{1.0, 2.0}
    shared_ptr<double[6][2]> sp10 = make_shared<double[6][2]>({1.0, 2.0});
    //std::shared_ptr 指向一个向量数组 vector<int>[4]，每一向量都是{5, 6}
    shared_ptr<vector<int>[4]>sp11 = make_shared<vector<int>[4]>({5, 6});
    return 0;
}
```

例 7-15 介绍了使用 std::make_shared 创建指向多个数据类型的一组对象（即数组）的 std::shared_ptr<T[]>指针的方法。请注意，在创建 std::shared_ptr 指针用于指向数组时，使用了两种不同的方法：指定数组的大小（如 std::shared_ptr<T[10]>）和不指定数组的大小（如 std::shared_ptr<T[]>）。std::make_shared 的第一个参数可用来确定数组的大小，后面的数据可用于初始化数组。

另外一点需要牢记的是，使用 std::shared_ptr<T[]>或者 std::unique_ptr<T[]>指向数组对象时，并不是创建了 STL 的容器，不能像使用普通容器那样直接使用 STL 的算法库函数。请看下面的代码：

```
const size_t COUNT = 10;
std::unique_ptr<int[]> ptr = std::make_unique<int[]>(10);
std::sort(ptr.begin(), ptr.end());              //编译错误：ptr 不是容器
std::sort(ptr.get(),   ptr.get() + COUNT);      //正确
auto ptr = std::make_unique<int[]>(COUNT);
std::iota(ptr.get(), ptr.get() + COUNT, 0);
```

表 7.4-2 比较了 std::unique_ptr<T[ ]>与 std::array、std::vector 的用法。

表 7.4-2 std::unique_ptr<T[ ]>与 std::array、std::vector 的用法比较

用法	std::array	std::vector	std::unique_ptr<T[ ]>
初始容量	在编译时确定	运行时确定	运行时确定
动态调整大小	不可以	可以	不可以
存储空间	数据存储在栈中	数据存储在堆中	数据存储在堆中
拷贝	容许拷贝	容许拷贝	不容许拷贝
交换/移动语义	时间复杂度为 $O(n)$	时间复杂度为 $O(1)$	时间复杂度为 $O(1)$
指针/迭代器失效	无	有	无
兼容性	普通容器类	普通容器类	非容器类

std::shared_ptr 的成员函数据绝大多数和 std::uniqu_ptr 相同，我们不再举例介绍。下面的示例

主要使用 std::shared_ptr 中特有的成员函数。

【例 7-16】

```cpp
#include <iostream>
#include <memory>
using namespace std;

class A {
public:
    void show(){ cout << "A::show()" << endl;}
};

int main(){
    shared_ptr<A> p1(new A);
    cout << p1.get() << endl;           //打印 p1 管理对象的地址
    p1->show();

    shared_ptr<A> p2(p1);
    p2->show();

    cout << p1.get() << endl;
    cout << p2.get() << endl;

    cout << p1.use_count() << endl;
    cout << p2.use_count() << endl;

    p1.reset();
    cout << p1.get() << endl;
    cout << p2.use_count() << endl;
    cout << p2.get() << endl;
    return 0;
}
```

在例 7-16 中，p1.get() 和 p2.get() 都将返回存储对象的地址，并且由于两者管理的是同一对象，所以输出的地址值是相同的。但在 p1.reset() 后，p1.get() 将返回 nullptr；p2 管理的仍然是对象 A，所以 p2.use_count() 将返回引用数 1。

例 7-17 提供了一个智能指针的上行转换（upcasting）和下行转换（downcasting）样例。上行转换通常是指将派生类转换成基类，下行转换则是指将基类转换成派生类。在面向对象的开发中经常使用这两种转换。例 7-17 实现了两个模板函数：static_pointer_cast 和 dynamic_pointer_cast，它们负责将一个类型转换成另一个类型。在本例中，在基类 Base 中定义了一个虚函数 virtual void f() const，在派生类 Derived 中对该虚函数重新定义。override 的作用是，在定义 Derived 的派生类中，如果使用的是 void f()const，则仍然可以重新定义虚函数，以呈现多态性；如果使用的是 void f()const final，则表示虚函数是最终的实现，在派生类中不能再重新定义了。static_pointer_cast 和 dynamic_pointer_cast 的定义为：

```cpp
template<class T, class U>
std::shared_ptr<T> static_pointer_cast(const std::shared_ptr<U>& r) noexcept
{
```

```cpp
        auto p = static_cast<typename std::shared_ptr<T>::element_type*>(r.get());
        return std::shared_ptr<T>{r, p};
}

template<class T, class U>
std::shared_ptr<T> dynamic_pointer_cast(const std::shared_ptr<U>& r) noexcept
{
        if(auto p = dynamic_cast<typenamestd::shared_ptr<T>::element_type*>(r.get()))
        return std::shared_ptr<T>{r, p};
        else
        return std::shared_ptr<T>{};
}
```

【例 7-17】

```cpp
#include <iostream>
#include <memory>

class Base{
public:
    int a;
    virtual void f() const { std::cout << "I am base!\n"; }
    virtual ~Base() {}
};

class Derived : public Base{
public:
    void f() const override { std::cout << "I am derived!\n";}
    ~Derived() {}
};

int main(){
    auto basePtr = std::make_shared<Base>();
    std::cout << "Base pointer says: ";
    basePtr->f(); //打印：I am base!

    auto derivedPtr = std::make_shared<Derived>();
    std::cout << "Derived pointer says: ";
    derivedPtr->f(); //打印：I am derived!

    //类型转换，从派生类到基类
    basePtr = std::static_pointer_cast<Base>(derivedPtr);
    std::cout << "Base pointer to derived says: ";
    basePtr->f(); //打印：I am derived!

    //类型转换，从基类到派生类
    auto downcastedPtr = std::dynamic_pointer_cast<Derived>(basePtr);
    if (downcastedPtr){
        std::cout << "Downcasted pointer says: ";
        downcastedPtr->f();    //打印：I am derived!
```

```
        }
        std::cout << "Pointers to underlying derived: "
                 << derivedPtr.use_count()    //打印：3
                 << '\n';
}
```

下面我们看看 std::shared_ptr 的内部工作机理。

当使用 std::shared_ptr 创建动态被管理对象时，std::shared_ptr 对象包含两个原生指针（raw pointer），第一个原生指针指向被管理对象，第二个原生指针指向一个控制块。控制块也包含一个原生指针（指向被管理对象）和两个引用计数器（一个引用计数器是被管理对象被 std::shared_ptr 引用的数量，另一个引用计数器是被管理对象被 std::weak_ptr 引用的数量）。根据初始化选项不同，控制块中可能还包含其他数据，如 deleter。shared_ptr 对象本身是在栈中创建的，但其管理的对象（图 7.4-1 中的 myClass）和控制块是在堆中创建的，并且是在 std::shared_ptr 对象的构造函数中创建被管理对象的。这样当程序运行至 std::shared_ptr 的范围之外时，将触发调用 std::shared_ptr 对象的析构函数，在析构函数中，被管理对象被销毁，动态分配的内存被释放。

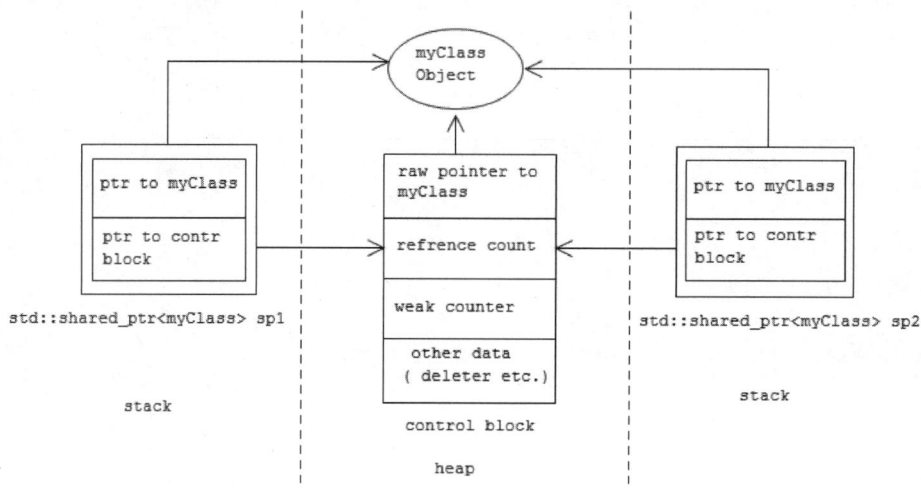

图 7.4-1 shared_ptr 工作机理

std::shared_ptr 可能会造成回环引用问题。在下面的例子里，我们看到 MyClassA 中有 std::shared_ptr<MyClassB> b，而 MyClassB 中有 std::shared_ptr<MyClassB> a，构成了你中有我、我中有你的形式，因而被管理对象不能被销毁，造成内存泄漏。

【例 7-18】

```
#include <memory>
#include <iostream>
struct MyClassB;
struct MyClassA{
    std::shared_ptr<MyClassB> b;
    ~MyClassA() { std::cout << "MyClassA::~MyClassA()\n"; }
};
struct MyClassB{
    std::shared_ptr<MyClassA> a;
    ~MyClassB() { std::cout << "MyClassB::~MyClassB()\n"; }
};
```

```cpp
void useClassAnClassB(){
    auto a = std::make_shared<MyClassA>();
    auto b = std::make_shared<MyClassB>();
    a->b = b;
    b->a = a;
}
int main(){
    useClassAnClassB();
    std::cout << "Finished using A and B\n";
    return 0;
}
```

7.4.3　std::weak_ptr

std::weak_ptr 作为一种智能指针，不能独立管理内存对象，并且不能直接使用 new 进行初始化，也没有与之对应的 std::make_weak，只能与 std::shared_ptr 一起使用。因此，std::weak_ptr 不在对象的引用计数器中，其成员函数 use_count() 返回被管理对象被 std::shared_ptr 引用的数量。

在 C++11 中，std::weak_ptr 是一个模板类，其定义如下：

```
class<T>
class    weak_ptr
```

【例 7-19】

```cpp
#include <iostream>
#include <memory>
using namespace std;
class Circle {
    double x;
    double y;
    double radius;
public:
    Circle(double x0, double y0, double r): x(x0),y(y0), radius(r){}
    double area(){ return PI*r*r;}
    double perimeter() {return 2.0*PI*r;}
};
int main(){
    shared_ptr<Circle> ptr1 = make_shared<Circle>(5.0, 5.0, 10.0);
    weak_ptr<Object> ptr2 = ptr1;

    if (!ptr2.expired()) {
        cout << "The area and perimeter of the circle is: "<< (*ptr2.lock()).area() << ","
            << (*ptr2.lock()).perimeter() << endl;
    }
    ptr1.reset();                           //删除对象
    cout << "End of the Program";
    return 0;
}
```

在例 7-19 中，成员函数 expired() 用来检查被管理对象是否已经被删除，成员函数 lock() 将返回一个 std::shared_ptr 对象，成员函数 reset() 则彻底删除被管理对象。

【例 7-20】

```cpp
#include <iostream>
#include <memory>

void observe(std::weak_ptr<int> weak){
    if (auto p = weak.lock())
        std::cout << "\tobserve() is able to lock weak_ptr<>, value=" << *p << '\n';
    else
        std::cout << "\tobserve() is unable to lock weak_ptr<>\n";
}

int main(){
    std::weak_ptr<int> weak;
    std::cout << "weak_ptr<> is not yet initialized\n";
    observe(weak);
    {
        auto shared = std::make_shared<int>(42);
        weak = shared;
        std::cout << "weak_ptr<> is initialized with shared_ptr\n";
        observe(weak);
    }
    std::cout << "shared_ptr<> has been destructed as scope exit\n";
    observe(weak);
    return 0;
}
```

另外，std::weak_ptr 的重要作用是它很好地解决了 std::shared_ptr 回环引用问题，我们对例 7-20 稍微做了修改，在 struct MyClassB 中定义 std::weak_ptr<MyClassA> a。

【例 7-21】

```cpp
#include <memory>
#include <iostream>

struct MyClassB;
struct MyClassA{
    std::shared_ptr<MyClassB> b;
    ~MyClassA() { std::cout << "MyClassA::~MyClassA()\n"; }
};
struct MyClassB{
    std::weak_ptr<MyClassA> a;   //使用 std::weak_ptr
    ~MyClassB() { std::cout << "MyClassB::~MyClassB()\n"; }
};
void useMyClassAnMyClassB(){
    auto a = std::make_shared<MyClassA>();
    auto b = std::make_shared<MyClassB>();
    a->b = b;
```

```
        b->a = a;
    }
    int main(){
        useMyClassAnMyClassB();
        std::cout << "Finished using A and B\n";
        return 0;
    }
```

7.5 本章小结

在 C++程序开发中，内存作为一种资源，分为栈和堆两种。栈的规模比堆要小。栈是函数参数、函数内临时变量的存储空间。一旦函数运行结束，所有的函数参数和函数内临时变量所占用的空间都将被释放。栈的特点是效率高、速度快。堆是程序员通过特殊指令创建动态对象时的存储空间总称。堆的规模大，但访问速度比栈慢。

使用指针可以大大提高程序开发效率，但也有不小的风险。内存泄漏是指针造成的最严重问题之一，现代 C++通过智能指针成功避免了这一难题。智能指针是模板类，将被引用的对象封装在其中，模板类是被引用对象的类型。

分段错误是当程序尝试访问它无权访问的内存空间时发生的错误。使用智能指针是有效解决问题的方法之一。

std::unique_ptr 是独占式智能指针，独自管理被引用的对象。虽然被引用的对象是在堆中分配存储空间的，当 std::unique_ptr 超出其定义的作用范围时，将触发被引用对象销毁。

std::shared_ptr 是共享式智能指针，多个 std::shared_ptr 指针可以引用一个共同的对象，但只有当最后一个指针被释放时，所引用的对象才会被销毁。

使用 std::shared_ptr 可避免回环引用问题，即避免甲引用乙、乙引用丙、丙引用甲的形式，在复杂的应用开发中，当回环引用不可避免时，请使用 std::weak_ptr。

在声明 std::unique_ptr 和 std::shared_ptr 时，可以使用 std::make_unique 和 std::make_shared 创建被引用对象，它们是模板函数，使用方便，这样可以完全不使用 new 和 delete。

std::weak_ptr 的重要用途之一是避免 std::shared_ptr 的回环引用问题。std::weak_ptr 只能与 std::shared_ptr 一起使用，不能单独使用。

第 8 章 并发与多线程

8.1 并发与并行

并发（concurrent）是指程序在同一时间段内处理多个任务，是利用单个处理器（CPU）提高系统响应速度的一种方法。并发制造了并行上的幻觉，但任务本身不是并行处理的。并发的实现是通过 CPU 进程调度，即 CPU 为每一个任务分配一个时间片，并通过上下文切换实现的。并发强调的是在同一时间段内处理多个任务。

并行（parallel）是指一个应用被分割成多个子任务，每个子任务在一个独立的处理器（CPU）上运行，或者在独立的计算机上运行，从而形成了并行处理。这些计算机物理上可能是异地的，构成了一个分布式处理系统。并行的关键是通过使用多处理器提高整个系统的处理速度和吞吐能力。

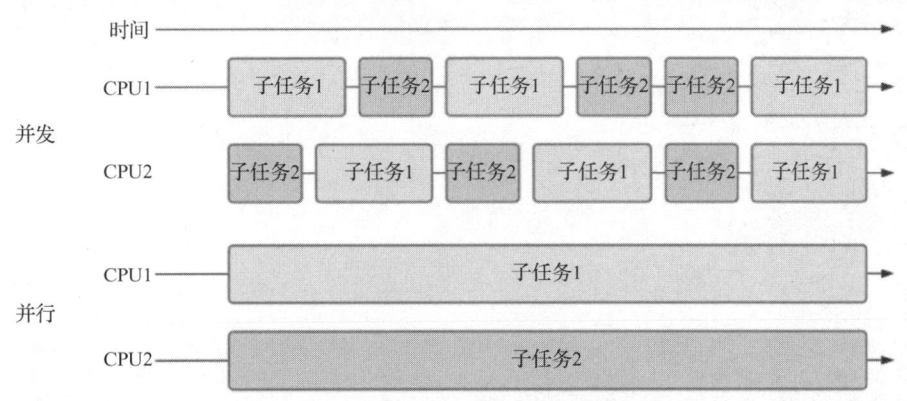

图 8.1-1 并发与并行的示意图

本章主要介绍并发与多线程，第 9 章将介绍并行算法与并行计算。下面给出了几个有助于了解并发和并行的几个概念：

（1）多处理（multiprocessing）：在单一的计算机系统中部署两个或者两个以上的 CPU 称为多处理。

（2）多线程（multithreading）：这是一种技术，可以让单个进程有多个代码段，如线程，这些代码段根据进程的上下文并发地运行。

（3）分布式计算（distributed computing）：一个分布式计算系统由多台计算机系统组成，并整体上像单一的系统运行。分布式系统中的计算机在地理布置上非常近并通过本地网络连接起来，或者在地理位置上非常远并通过广域网连接起来。

（4）多核处理器（multicore processor）：它是一个集成处理器，包含多个核处理单元，又称为 CMP（chip multiprocessor）。

（5）管道技术（pipelining）：这是一种技术，可以让多个指令在单次执行中叠加在一起。

8.2 创建线程

并发是通过多线程实现的。C++11 引入了 std::thread 类来实现多线程编程。在 C++11 之前，UNIX/Linux 操作系统支持使用 POSIX threads 或<pthreads>库进行多线程编程。本章不再回顾传统 C++中有关多线程的 API 函数，直接介绍 C++11 的 std::thread 类。std::thread 线程类及相关的函数放在头文件<thread>中。std::thread 的语法为：

```
std::thread thread_object (callable);
```

要启动一个线程，首先需要定义 std::thread 线程类的一个对象，并将要执行的函数传入线程对象的构造函数。如果要启动多个线程，就需要定义多个线程类的实例，从而让这些线程并发运行。线程类的实例一生成，该线程就启动了。这里的可调用对象可以是函数指针、lambda 表达式、函数对象、非静态成员函数、静态成员函数。

下面给出了使用不同可调用对象创建线程的示例。

【例 8-1】

```cpp
#include <chrono>
#include <iostream>
#include <thread>
#include <utility>
void f1(int k){
    for (int i = 0; i < 5; ++i){
        std::cout << "Thread 1 is running...\n";
        ++k;
        std::this_thread::sleep_for(std::chrono::milliseconds(15));
    }
}

void f2(int& k){
    for (int i = 0; i < 5; ++i){
        std::cout << "Thread 2 is running...\n";
        ++k;
        std::this_thread::sleep_for(std::chrono::milliseconds(15));
    }
}

int main(){
    int n = 0;
    std::thread t1(f1, n + 1);               //传值方式
    std::thread t2(f2, std::ref(n));         //引用方式
    t1.join();
    t2.join();
    std::cout << "n = " << n << endl;
}
```

【例 8-2】

```cpp
#include <iostream>
#include <thread>
int main(){
    //定义一个 lambda 表达式
    auto f = [](int n) {
        for (int i = 0; i < n; i++)
            std::cout << "Thread 3 is running...\n";
    };

    thread t(f, 5);
    t.join();
    return 0;
}
```

【例 8-3】

```cpp
#include <chrono>
#include <iostream>
#include <thread>
#include <utility>

class myObject{
public:
    void operator()(int m){
        for (int i = 0; i < m; ++i){
            std::cout << "Thread 4 is running\n";
            ++n;
            std::this_thread::sleep_for(std::chrono::milliseconds(10));
        }
    }
    int n = 0;
};

int main(){
    myObject obj;
    std::thread t(obj(), 5);
    t.join();
    std::cout <<"obj.n (obj::n) is " << obj.n << endl;
    return 0;
}
```

【例 8-4】

```cpp
#include <chrono>
#include <iostream>
#include <thread>
#include <utility>

class myObject{
```

```cpp
public:
    void func(){
        for (int i = 0; i < 5; ++i){
            std::cout << "Thread 5 is running\n";
            ++n;
            std::this_thread::sleep_for(std::chrono::milliseconds(10));
        }
    }
    int n = 0;
};

int main(){
    myObject obj;
    std::thread t(&myObject::func, &obj);
    t.join();
    std::cout << "obj.n (myObject::n) = " << obj.n << endl;
}
```

【例 8-5】

```cpp
#include <iostream>
#include <thread>

class myObject {
public:
    static void func(){
        cout << "Thread 6 is running\n" << endl;
    }
};

int main(){
    thread t(&myObject::func);
    t.join();
    return 0;
}
```

我们已经知道如何创建并启动一个线程，下面我们看看 std::thread 类有哪些成员函数，并通过示例介绍这些成员函数的使用方法。std::thread 的成员函数如表 8.2-1 所示。

表 8.2-1 std::thread 的成员函数

成 员 函 数	说　　明
operator=	线程对象赋值
joinable	检查线程是否可能在并行上下文中运行
get_id	返回线程的 ID
native_handle	返回当前实现定义的线程句柄
hardware_concurrency	返回实现支持的并发线程数
join	等待线程执行完成
detach	允许线程独立于线程句柄执行
swap	交换两个线程对象

【例 8-6】

```cpp
#include <chrono>
#include <iostream>
#include <thread>

void f1(){
    std::this_thread::sleep_for(std::chrono::seconds(10));
}
void f2(){
    std::this_thread::sleep_for(std::chrono::seconds(10));
}

int main(){
    std::thread t1(f1);
    std::thread t2(f2);
    std::cout << "thread 1 id: " << t1.get_id() << '\n'
              << "thread 2 id: " << t2.get_id() << '\n';
    std::swap(t1, t2);
    std::cout << "after std::swap(t1, t2):" << '\n'
              << "thread 1 id: " << t1.get_id() << '\n'
              << "thread 2 id: " << t2.get_id() << '\n';
    t1.swap(t2);
    std::cout << "after t1.swap(t2):" << '\n'
              << "thread 1 id: " << t1.get_id() << '\n'
              << "thread 2 id: " << t2.get_id() << '\n';
    t1.join();
    t2.join();
    return 0;
}
```

【例 8-7】

```cpp
#include <chrono>
#include <iostream>
#include <thread>

void independentThread() {
    std::cout << "Starting concurrent thread.\n";
    std::this_thread::sleep_for(std::chrono::seconds(5));
    std::cout << "Exiting concurrent thread.\n";
}

void threadCaller() {
    std::cout << "Starting thread caller.\n";
    std::thread t(independentThread);
    t.detach();
    std::this_thread::sleep_for(std::chrono::seconds(5));
    std::cout << "Exiting thread caller.\n";
}
```

```cpp
int main(){
    threadCaller();
    std::this_thread::sleep_for(std::chrono::seconds(5));
    return 0;
}
```

【例 8-8】

```cpp
#include <chrono>
#include <iostream>
#include <thread>

void f(){
    std::this_thread::sleep_for(std::chrono::seconds(5));
}

int main(){
    std::thread t1(f);
    std::thread::id t1_id = t1.get_id();
    std::thread t2(f);
    std::thread::id t2_id = t2.get_id();
    std::cout << "t1's id: " << t1_id << '\n';
    std::cout << "t2's id: " << t2_id << '\n';

    t1.join();
    t2.join();

    std::cout << "t1's id after join: " << t1.get_id() << '\n';
    std::cout << "t2's id after join: " << t2.get_id() << '\n';
    return 0;
}
```

8.3 线程同步与互斥

在多线程编程中，线程同步（thread synchronization）是一个非常重要且必须考虑的问题。线程同步保证了整个程序平稳运行，产生可预测的结果。线程同步指的是多个线程之间协调和合作，以确保这些线程在特定时刻的执行顺序、时间或者条件达成。线程同步的目的是保证多个线程能够在正确的时间点上相互配合，以完成某项任务。线程互斥（Thread Mutual Exclusion）指的是通过使用互斥锁等机制，来防止多个线程同时访问临界资源，从而避免数据竞争和不一致性问题。互斥锁是一种常用的同步机制，用于保护共享资源，确保在任何时刻只有一个线程能够访问共享资源。

8.3.1 std::mutex

std::mutex 是一种互斥锁，用来防止共享变量被多个线程同时修改。std::mutex 采用一种排他的、非递归的占有机制。线程通过使用 mutex.lock()、mutex.try_lock()和 mutex.unlock()来实现多线程下对共享变量的访问。注意，std::mutex 是在头文件<mutex>中定义的。

【例 8-9】

```cpp
#include <chrono>
#include <iostream>
#include <mutex>
#include <thread>
int counter = 0;
std::mutex counter_mutex;

void counter_func(int id){
    for (int i = 0; i < 10; ++i)
    {
        counter_mutex.lock();                       //加锁
        ++counter;
        std::cout << "id: " << id << ", counter: " << counter << '\n';
        counter_mutex.unlock();                     //解锁
        std::this_thread::sleep_for(std::chrono::milliseconds(234));
    }
}

int main(){
    std::thread t1{counter_func, 0};
    std::thread t2{counter_func, 1};
    t1.join();
    t2.join();
    return 0;
}
```

8.3.2　std::condition_variable

std::condition_variable 称为条件变量，是一种同步机制，提供 wait()、wait_for()和 wait_until()成员函数来堵塞当前线程；同时，通过 notify_one()通知另一个线程和通过 notify_all()通知所有线程。std::condition_variable 与 std::mutex 结合使用，可堵塞一个或多个线程，直到线程完成对共享资源的修改并且通知 std::condition_variable 为止。使用 std::condition_variable 时，需要在程序中添加"include <condition_variable>"。

【例 8-10】

```cpp
#include <condition_variable>
#include <iostream>
#include <mutex>
#include <thread>
std::mutex mtx;
std::condition_variable cv;                         //声明 cv
bool data_ready = false;                            //全局变量

void producer_thread(){
```

```cpp
        std::this_thread::sleep_for(std::chrono::seconds(2));
        std::lock_guard<mutex> lock(mtx);
        data_ready = true;                      //修改 data_ready 的值
        std::cout << "Data Produced!" << endl;
        cv.notify_one();                        //通知其他线程
}

void consumer_thread(){
    std::unique_lock<mutex> lock(mtx);
    cv.wait(lock, [] { return data_ready; });   //堵塞线程
    std::cout << "Data consumed!" << endl;
}

int main(){
    std::thread consumer(consumer_thread);
    std::thread producer(producer_thread);
    consumer.join();
    producer.join();
    return 0;
}
```

8.3.3　std::lock_guard 和 std::unique_lock

在 C++11 中，我们一般不直接使用互斥量 std::mutex 进行加锁和解锁，而是为了遵循 RAII（资源获取即初始化）原则，使用 std::lock_guard 和 std::unique_lock（它们是 Mutex 类型的模板类，提供了一种更加方便的形式使用 std::mutex）。std::lock_guard 和 std::unique_lock 的语法如下：

```cpp
template< class Mutex >
class lock_guard;

template< class Mutex >
class unique_lock;
```

std::unique_lock 是 std::lock_guard 的超级形式，更加灵活但也更重量级（因为它需要管理更多的状态并支持移动语义），提供了更多的控制和更复杂的锁定机制，如延迟锁定、尝试锁定和定时锁定等。std::unique_lock 还拥有成员函数 lock()、try_lock()、try_lock_for()、try_lock_until()和 unlock()。下面通过多个例子来介绍 std::lock_guard，特别是 std::unique_lock 的使用方法。

【例 8-11】

```cpp
#include <chrono>
#include <iostream>
#include <map>
#include <mutex>
#include <string>
#include <thread>

std::map<std::string, std::string> g_data;     //需要访问受保护的全局数据
std::mutex g_data_mutex;
```

```cpp
void save_data(const std::string& key){
    std::this_thread::sleep_for(std::chrono::seconds(2));
    std::string result = "modern C++ tutorial";
    std::lock_guard <std::mutex> guard(g_data_mutex);
    g_data[key] = result;
}

int main()
{
    std::thread t1(save_data, "foo");
    std::thread t2(save_data, "bar");
    t1.join();
    t2.join();

    //安全访问 g_data 而不需要加锁，因为线程 t1 和 t2 已经结束
    for (const auto& pair:g_data)
    std::cout << pair.first << " => " << pair.second << '\n';
    return 0;
}
```

例 8-11 创建了两个线程 t1 和 t2，它们的线程函数是 save_data()，访问同一个全局变量 g_data，因此需要使用加锁机制。我们使用 std::mutex 并定义一个 std::mutex 类型的全局变量 g_data_mutex（互斥量），将该变量传入 std::lock_guard 的构造函数中，实现对全局数据变量 g_data 的加锁保护。需要记住：std::lock_guard 在构造实例后会立即锁住互斥量 g_data_mutex，直到超出活动范围后，std::lock_guard 才调用其析构函数来解锁互斥量。

std::unique_lock 在创建对象时是否立即上锁取决于它的构造函数的调用方式。std::unique_lock 提供了多种构造函数，允许在创建对象时设置是否立即上锁。另外，std::unique_lock 提供了其他高级功能。下面通过示例介绍 std::unique_lock 几种锁定方式。

第一种锁定方式是默认锁定方式，即在构造 std::unique_lock 时不指定任何加锁方式，在 std::unique_lock 创建对象时立即调用成员函数 lock() 来锁住互斥量。

【例 8-12】

```cpp
#include <chrono>
#include <iostream>
#include <mutex>
#include <thread>
#include <vector>

int main(){
    int counter = 0;
    std::mutex counter_mutex;
    std::vector<std::thread> threads;                    //存放线程的 vector

    auto worker_task = [&](int id)
    {
        std::unique_lock<std::mutex> lock(counter_mutex);    //立即锁住
        ++counter;
        std::cout << id << ", initial counter: " << counter << '\n';
```

```cpp
            lock.unlock();                                          //解锁

            std::this_thread::sleep_for(std::chrono::seconds(1));
            lock.lock();                                            //再次锁住
            ++counter;
            std::cout << id << ", final counter: " << counter << '\n';
        };                                                          //lock 自动解锁

        for (int i = 0; i < 10; ++i)                                //创建线程
            threads.emplace_back(thread(worker_task, i));

        for (auto& thread : threads)
            thread.join();
        return 0;
}
```

第二种锁定方式使用的构造函数是 std::defer_lock。std::unique_lock 在创建对象时延迟锁定互斥量，直到需要时再锁定。也就是说，创建 std::unique_lock 对象后，没有调用 lock()。

【例 8-13】

```cpp
#include <iostream>
#include <mutex>
#include <thread>

struct bank_account{
    explicit bank_account(int balance) : balance{balance}{}
    int balance;
    std::mutex m;
};

void transfer(bank_account & from, bank_account & to, int amount){
    if (&from == &to)                   //避免自己给自己转账时死锁
    return;
    //创建两个 std::defer_lock 的锁：lock1 和 lock2.
    std::unique_lock<std::mutex>lock1{from.m, std::defer_lock};
    std::unique_lock<std::mutex>lock2{to.m, std::defer_lock};
    //这里可以做一些其他事情
    std::lock(lock1, lock2);            //锁住两个互斥量，避免陷入锁死僵局
    from.balance -= amount;
    to.balance += amount;
}

int main(){
    bank_account my_account{100};
    bank_account your_account{50};
    std::thread t1{transfer, std::ref(my_account), std::ref(your_account),10};
    std::thread t2{transfer, std::ref(your_account), std::ref(my_account),5};
    t1.join();
    t2.join();
```

```cpp
        std::cout<<"my_account.balance="<<my_account.balance <<"\n"
                <<"your_account.balance="<<your_account.balance<<"\n";
    return 0;
}
```

第三种锁定方式使用的构造函数是 std::try_to_lock。std::unique_lock 在创建对象时指定加锁方式，std::try_to_lock 在 std::unique_lock 创建对象后调用它的成员函数 try_lock()而不是 lock()。

【例 8-14】

```cpp
#include <iostream>
#include <mutex>
#include <thread>
#include <queue>
std::queue <int> m_que;
std::mutex m_utex;

void consumer(){
    int i=0, ncount=0;
    while(i<10){
        std::unique_lock<std::mutex> lk(m_utex,std::try_to_lock);
        if(lk.owns_lock()){
            if(m_que.empty()){
                char buf[48] = {0};
                sprintf_s(buf,"consumer fetch out %d", m_que.front());
                cout << buf;
                m_que.pop();
                i++;
            }
        }
        else{
            ncount++;
            char buf[48] = {0};
            sprintf_s(buf, "consumer has tried %d times", ncount);
            cout << buf;
        }
    }
}

void producer(){
    int i=0;
    while( i<10){
        std::this_thread::sleep_for(std::chrono::milliseconds(200));
        std::lock_guard<std::mutex> lc(m_utex);            //立即加锁
        char buf[48] = {0};
        sprintf_s(buf, "producer insert %d into m_que", i+1);
        m_que.push(i+1);
        cout << buf;
        i++;
    }
```

```cpp
}
int main(){
    std::thread th1(producer);
    std::thread th2(consumer);
    th1.join();
    th2.join();
    return 0;
}
```

第四种锁定方式使用的构造函数是 std::adopt_lock。当互斥量已经被手动锁定后,在 std::unique_lock 或者 std::lock_guard 创建对象时可通过 std::adopt_lock 来接管互斥量的所有权,这等于告诉 std::unique_lock 创建的对象,互斥量已经被锁定。

【例8-15】

```cpp
#include <iostream>
#include <mutex>
#include <thread>

struct bank_account{
    explicit bank_account(int balance) : balance{balance} {}
    int balance;
    std::mutex m;
};

void transfer(bank_account& from, bank_account& to, int amount){
    if (&from == &to)                                    //避免自己给自己转账出现死锁
        return;
    std::lock(from.m, to.m);                             //两个互斥锁已经被锁定
    std::lock_guard lock1{from.m, std::adopt_lock};      //获取互斥锁的所有权
    std::lock_guard lock2{to.m, std::adopt_lock};        //获取互斥锁的所有权
    from.balance -= amount;
    to.balance += amount;
}

int main(){
    bank_account account_1{80};
    bank_account account_2{80};
    std::thread t1{transfer, std::ref(account_1),
        std::ref(account_2), 20};
    std::thread t2{transfer, std::ref(account_2),
        std::ref(account_1), 30};
    t1.join();
    t2.join();
    std::cout << "account_1.balance=" << account_1.balance <<"\n";
    std::cout << "account_2.balance=" << account_2.balance <<"\n";
    return 0;
}
```

8.3.4　std::atomic

C++11 引入了 std::atomic，用于管理多线程应用程序对共享资源的并发访问。std::atomic 提供原子类型和操作，可确保安全访问共享资源，防止多线程应用程序中的数据争用和潜在问题。std::atomic 通过原子操作来允许线程安全地访问和修改共享资源，而无须显式加锁或同步机制。在原子操作中，并发数据访问由内存模型调节，如果一个线程尝试访问另一个线程正在访问的数据，则该模型的行为是明确定义的。std::atomic 是一个类模板，其语法如下：

```
template< class T >
struct atomic;

template< class U >
struct atomic<U*>;
```

std::atomic 的成员函数如表 8.3-1 所示。

表 8.3-1　std::atomic 的成员函数

成员函数	说明
load()	该函数将值加载到原子对象中
store()	该函数将值存储在原子对象中
exchange()	该函数替换存储在原子对象中的值，并返回先前存储的值
wait()	该函数用于阻止线程
notify_one()	通知正在等待原子对象的一个线程
notify_all()	通知正在等待原子对象的所有线程
fetch_add()	获取存储的当前值，并将当前值和原子对象的值相加
fetch_sub()	获取存储的当前值，并从原子对象的值中减去当前值
fetch_and()	获取存储的当前值，对原子对象中的值和当前值进行按位与运算
fetch_or()	获取存储的当前值，对原子对象中的值和当前值进行按位或运算
fetch_xor()	获取存储的当前值，对原子对象中的值和当前值进行按位异或运算

【例 8-16】

```
#include <atomic>
#include <iostream>
#include <thread>

std::atomic<int> counter{0};                    //创建原子对象 counter

void increment(int id) {
    for (int i = 0; i < 100000; ++i) {
        counter.fetch_add(1);                   //获取原子对象的当前值，并加 1
    }
}

int main() {
```

```cpp
    std::thread th0(increment, 1);
    Std::thread th1(increment, 2);

    th0.join();
    th1.join();

    //counter.load()加载原子对象中的值
    std::cout << "Counter: " << counter.load() << std::endl;
    return 0;
}
```

【例 8-17】

```cpp
#include <iostream>
#include <atomic>
#include <thread>
#include <vector>

std::atomic<bool> ready (false);           //原子对象 ready 被初始化为 false
std::atomic<bool> winner (false);          //原子对象 winner 被初始化为 false

void counter (int id) {
    while (!ready) {}                      //等待 ready 变为 true
    for (int i=0; i<1000000; ++i) {}       //计数到 100 万
    if (!winner.exchange(true)) {          //如果原子对象中的值为 false，则替换为 true
        std::cout << "thread #" << id << " won!\n";
    }
};

int main (){
    std::vector<std::thread> threads;
    std::cout << "spawning 10 threads that count to 1 million...\n";
    for (int i=1; i<=10; ++i) threads.push_back(std::thread(counter,i));
    ready = true; //ready 置为 true
    for (auto& th : threads) th.join();
    return 0;
}
```

8.4 线程死锁

8.4.1 std::lock

在 C++编程中，我们经常需要一次性操作多个临界资源，并且这些临界资源的锁是独立的（每个临界资源对应一把锁）。当多个线程或者进程对这些资源进行并发访问时，如果不注意锁的顺序就会出现死锁（deadlock）问题。std::lock 就是用来解决死锁问题的。注意：std::lock 是一个模板函数，不是类模板，定义在头文件<mutex>中，用于加锁所有的对象、堵塞调用的线程，并以未指定的顺序调用 std::mutex 的成员函数，如 lock()、try_lock()、unlock()。std::lock 的语法为：

```
template < class Lockable1, class Lockable2, class... LockableN >
void lock(Lockable1& lock1, Lockable2& lock2, LockableN&... lockn);
```

这里的 lock1，lock2，…，lockn 是加锁对象。

【例 8-18】

```cpp
#include <iostream>
#include <thread>
#include <mutex>
std::mutex m1, m2;

void task_a(){
    std::lock (m1,m2);
    std::cout << "task a\n";
    m1.unlock();
    m2.unlock();
}

void task_b(){
    std::lock (m2,m1);
    std::cout << "task b\n";
    m2.unlock();
    m1.unlock();
}

int main (){
    std::thread th1 (task_a);
    std::thread th2 (task_b);
    th1.join();
    th2.join();
    return 0;
}
```

【例 8-19】

```cpp
#include <mutex>
#include <thread>
#include <iostream>
#include <vector>
#include <functional>
#include <chrono>

struct Employee{
    Employee(int id):id(id){}
    int id;
    std::vector<int> lunch_partners;
    std::mutex m;
};

void send_mail(Employee &e1, Employee &e2){
    std::this_thread::sleep_for(std::chrono::seconds(1));
```

```cpp
}
void assign_lunch_partner(Employee &e1, Employee &e2){
    std::lock(e1.m, e2.m);
    e1.lunch_partners.push_back(e2.id);
    e2.lunch_partners.push_back(e1.id);
    e1.m.unlock();
    e2.m.unlock();
    send_mail(e1, e2);
    send_mail(e2, e1);
}

int main(){
    Employee Alex(1), Brad(2), David(3), Jack(4), Mark(5), Tom(6)
    std::vector<std::thread> threads;
    threads.emplace_back(assign_lunch_partner,std::ref(Alex), std::ref(Brad));
    threads.emplace_back(assign_lunch_partner,std::ref(David), std::ref(Jack));
    threads.emplace_back(assign_lunch_partner,std::ref(Mark), std::ref(Tom));
    for (auto &thread : threads) thread.join();
    return 0;
}
```

将 std::lock、std::adopt_lock 和 std::lock_guard 结合起来可以有效解决死锁问题。例如，下面的代码是将三者结合使用的。

```cpp
struct resource{
    int n1{0};
    std::mutex mtx;
};
void swap(resource &first, resource &second){
    std::lock_guard<std::mutex> lock1(first.mtx);
    std::lock_guard<std::mutex> lock2(second.mtx);
    int temp = first.n1;
    first.n1 = second.n1;
    second.n1 = temp;
}
```

上面代码中的 swap() 函数用来交换 resource 的内部数据，在该函数内部会对传入的两个 resource 实例的锁进行依次上锁，然后进行数据交换。对于这两个实例中的锁进行上锁的先后顺序取决于实例的传入顺序。如果有两个线程同时对这两个实例执行数据交换操作，但实例传入的顺序是反的，则有死锁的风险。改进后的 swap() 函数如下：

```cpp
void swap(resource &first,resource &second){
    std::lock(first.mtx,second.mtx);
    std::lock_guard<std::mutex> lockf(first.mtx,std::adopt_lock);
    std::lock_guard<std::mutex> locks(second.mtx,std::adopt_lock);
    int temp = first.n1;
    first.n1 = second.n1;
    second.n1 = temp;
}
```

改进后的代码先通过 std::lock 对两个锁进行"同时"上锁，然后通过 std::lock_guard 和 std::adopt_lock 对两个锁进行了所有权的持有。但 std::lock_guard 在超越其活动范围之外时，会自动释放所有权，这样两个线程在交换 resource 的内部数据时不会死锁。

8.4.2　std::scoped_lock

std::scoped_lock 是一个 Mutex 类模板，可为 0 到多个 Mutex 对象加锁。当 std::scoped_lock 离开它的活动范围后，就调用自己的析构函数释放掉 Mutex 对象。

【例 8-20】

```cpp
#include <chrono>
#include <functional>
#include <iostream>
#include <mutex>
#include <string>
#include <thread>
#include <vector>
using namespace std;
using namespace std::chrono_literals;

struct Employee{
    vector<string> lunch_partners;
    string id;
    mutex m;
    Employee(string id) : id(id) {}
    string partners() const{
        string ret="Employee " + id + " has lunch partners:";
        for (int count{}; const auto& partner : lunch_partners)
            ret += (count++ ? ", " : "") + partner;
        return ret;
    }
};

void send_mail(Employee&, Employee&){
    this_thread::sleep_for(1s);
}

void assign_lunch_partner(Employee& e1, Employee& e2){
    static mutex io_mutex;
    {
        lock_guard<std::mutex> lk(io_mutex);
        cout <<e1.id<<"and"<<e2.id<<"are waiting for locks"<<endl;
    }
    {
        scoped_lock lock(e1.m, e2.m);
        {
            lock_guard<std::mutex> lk(io_mutex);
            cout<<e1.id<<" and "<<e2.id<<"got locks" << endl;
```

```cpp
            }
            e1.lunch_partners.push_back(e2.id);
            e2.lunch_partners.push_back(e1.id);
        }

        send_mail(e1, e2);
        send_mail(e2, e1);
}

int main(){
    Employee a("Alice"),b("Bob"),c("Christina"), d("Dave");
    std::vector<std::thread> threads;
    threads.emplace_back(assign_lunch_partner, ref(a), ref(b));
    threads.emplace_back(assign_lunch_partner, ref(c), ref(b));
    threads.emplace_back(assign_lunch_partner, ref(c), ref(a));
    threads.emplace_back(assign_lunch_partner, ref(d), ref(b));

    for (auto& thread : threads)
    thread.join();
    cout << a.partners() << '\n'    << b.partners() << '\n'
    << c.partners() << '\n' << d.partners() << '\n';
    return 0;
}
```

8.5 STL 中的<future>

<future>提供了一组类模板和函数模板来支持并发多线程编程，如异步访问由其他线程提供的变量。本节介绍<future>中几个主要的类和函数。

8.5.1 std::async

std::async 是一个异步调用函数，定义在头文件<future>中。std::async 能以异步和线程的方式运行一个可调用对象 fn。线程可以位于线程池中，并返回一个 std::future 对象，该对象包含了可调用对象 fn 的运行结果。std::async 的语法为：

```cpp
template<class Fn, class... Args>
future<typename result_of<Fn(Args...)>::type> async(std::launch policy, Fn&& fn, Args&&...args);
```

其中：

policy：是可调用对象 fn 的发起方式，主要发起方式有 std::launch::async（异步执行）、std::launch::deferred（同步执行）、std::launch::async | std::launch::deferred （由操作系统决定是异步执行还是同步执行）三种方式。如果 policy 项没有指定，则由操作系统决定是异步还是同步。

fn：可调用对象，可以是普通函数、lambda 表达式、类成员函数等。

args：可调用对象的参数列表。

返回值：std::async 返回一个 std::future 对象，它指向 std::async 的共享状态之一，deffered 表

示异步操作还没有开始，ready 表示异步操作已经完成，timeout 表示异步操作超时。

【例 8-21】

```cpp
#include <future>
#include <iostream>

void called_from_async() {
    std::cout << "Async call" << std::endl;
}
int main() {
    //在单独的线程中启动 called_from_async
    std::future<void> result ( std::async(called_from_async));
    std::cout << "Message from main." << std::endl;
    //如果 called_from_async 还没有启动，则确保它被同步启动
    result.get();
    return 0;
}
```

【例 8-22】

```cpp
#include <algorithm>
#include <future>
#include <iostream>
#include <mutex>
#include <numeric>
#include <string>
#include <vector>

std::mutex m;
struct X{
    void foo(int i, const std::string& str){
        std::lock_guard<std::mutex> lk(m);
        std::cout << str << ' ' << i << '\n';
    }
    void bar(const std::string& str){
        std::lock_guard<std::mutex> lk(m);
        std::cout << str << '\n';
    }
    int operator()(int i){
        std::lock_guard<std::mutex> lk(m);
        std::cout << i << '\n';
        return i + 10;
    }
};

template<typename RandomIt>
int parallel_sum(RandomIt beg, RandomIt end){
    auto len = end - beg;
    if (len < 1000)
        return std::accumulate(beg, end, 0);
```

```cpp
        RandomIt mid = beg + len / 2;
        auto handle = std::async(std::launch::async, parallel_sum<RandomIt>, mid, end);
        int sum = parallel_sum(beg, mid);
        return sum + handle.get();
}

int main(){
    std::vector<int> v(10000, 1);
    std::cout <<"The sum is "<<parallel_sum(v.begin(), v.end())<<'\n';

    X x;
    auto a1 = std::async(&X::foo, &x, 42, "Hello");
    auto a2 = std::async(std::launch::deferred, &X::bar, x, "world!");
    auto a3 = std::async(std::launch::async, X(), 43);
    a2.wait();
    std::cout << a3.get() << '\n';
    return 0;
}
```

8.5.2　std::future

std::future 是一个类模板，定义在头文件<future>中。std::future 提供了一种访问异步操作结果的机制。通常，一个异步操作是指创建一个 std::async、std::packaged_task 或 std::promise 对象，并且返回一个 std::future 对象。异步操作的对象可以通过 query、wait_for、extract 等成员函数从 std::future 中获取一个值。

表 8.5-1　std::future 的成员函数

成员函数	说明
operator=	移动 std::future 对象
share	将共享状态从*this 传输到 std::shared_future 并返回共享状态
get	返回结果
valid	检查未来是否具有共享状态
wait	等待结果变成可用
wait_for	等待结果，如果在指定的超时持续时间内不可用，则返回
wait_until	等待结果，如果在达到指定的时间点之前不可用，则返回

【例 8-23】

```cpp
#include <future>
#include <iostream>
#include <vector>

int twice(int m) {
    return 2 * m;
}
```

```cpp
int main() {
    std::vector<std::future<int>> futures;

    for(int i = 0; i < 10; ++i) {
        futures.push_back (std::async(twice, i));
    }

    //抽取并打印输出在 future 中的值
    for(auto &e : futures) {
        std::cout << e.get() << std::endl;
    }
    return 0;
}
```

【例 8-24】

```cpp
#include <future>
#include <iostream>
#include <thread>
int main(){
    //future from a packaged_task
    std::packaged_task<int()> task([]{ return 7; });
    std::future<int> f1 = task.get_future();
    std::thread t(std::move(task));
    //future from an async()
    std::future<int>f2 = std::async(std::launch::async,[]{return 8;});
    //future from a promise
    std::promise<int> p;
    std::future<int> f3 = p.get_future();
    std::thread([&p]{p.set_value_at_thread_exit(9); }).detach();
    std::cout << "Waiting..." << std::flush;
    f1.wait();
    f2.wait();
    f3.wait();
    std::cout <<"Done!\n Results are: " << f1.get() << ' ' << f2.get() << ' ' << f3.get() << '\n';
    t.join();
    return 0;
}
```

8.5.3 std::promise

std::promise 通常和 std::future 配合使用，用于两个线程间的通信，其作用是在一个线程中保存一个类型 T 的值，供绑定的 std::future 对象在另一线程中获取。通常 std::promise 用于发送线程端，std::future 用于接收线程端。std::promise 的成员函数如表 8.5-2 所示。

表 8.5-2 std::promise 的成员函数

成 员 函 数	说　　明
operator=	赋值共享状态

续表

成 员 函 数	说　　明
swap	交换两个 promise 对象
get_future	返回一个与 promised 关联的 future 结果
set_value	给结果设置特定的值
set_value_at_thread_exit	给结果设置特定的值，同时在离开线程时发出通知
set_exception	将结果设置为异常
set_exception_at_thread_exit	将结果设置为仅在线程出口时传递通知时指示异常

【例 8-25】

```cpp
#include <iostream>
#include <future>
#include <chrono>

void fn1(std::promise<int> &p){
    std::this_thread::sleep_for(std::chrono::seconds(5));
    int iVal = 210;
    std::cout << "sent(int): " << iVal << std::endl;
    p.set_value(iVal);
}

void fn2(std::future<int> &f){
    auto iVal = f.get();                            //iVal = 210
    std::cout << "received(int): " << iVal << std::endl;
}

int main(){
    //声明一个 std::promise 对象 pr1，其保存的值类型为 int
    std::promise<int> pr1;
    std::future<int> fu1 = pr1.get_future();

    std::thread t1(fn1, std::ref(pr1));
    std::thread t2(fn2, std::ref(fu1));
    t1.join();
    t2.join();
    return 0;
}
```

8.5.4　std::packaged_task

std::packaged_task 是一个类模板，其语法如下：

```cpp
template< class R, class ...ArgTypes >
class packaged_task <R(ArgTypes...)>;
```

其中，R 是可调用对象的返回数据类型，ArgTypes 是可调用对象的参数。

std::packaged_task 与 std::function 类似，可以绑定一个可调用对象（可调用对象可以是普通函数、lambda 表达式、类成员函数等），并使得被绑定的可调用对象被异步激活。std::packaged_task 的运行结果或者异常信息只能存储在一个共享状态里，只能通过 std::future 来访问。

为什么要使用 std::packaged_task 呢？我们回头看看用一个有返回值的函数 fn 来构造 std::thread 实例，结果发现函数的返回值被忽略了，无法获得。另外，如果函数 fn 出现异常，则 std::thread 会关闭整个程序。为了解决这个问题，C++11 的 STL 增加了类模板 std::packaged_task，并通过它的成员函数 get_future() 返回一个 std::future 对象（用于接收一个返回值）。std::packaged_task 还定义了成员函数 operator()，用来调用原始被绑定的可调用对象 fn。

【例 8-26】

```cpp
#include <future>
#include <iostream>

int compute(int a, int b) {
    return 42 + a + b;
}

int main() {
    std::packaged_task<int(int, int)> task(compute);
    std::future<int> f = task.get_future();
    task(3, 4);
    std::cout << f.get() << std::endl;
    return 0;
}
```

【例 8-27】

```cpp
#include <cmath>
#include <functional>
#include <future>
#include <iostream>
#include <thread>

int f(int x, int y){
    return std::pow(x, y);
}

void task_lambda(){
    std::packaged_task<int(int, int)> task([](int a, int b){
        return std::pow(a, b);
    });
    std::future<int> result = task.get_future();
    task(2, 9);
    std::cout << "task_lambda:\t" << result.get() << '\n';
}

void task_bind(){
    std::packaged_task<int()> task(std::bind(f, 2, 11));
    std::future<int> result = task.get_future();
    task();
```

```
        std::cout << "task_bind:\t" << result.get() << '\n';
}

void task_thread(){
    std::packaged_task<int(int, int)> task(f);
    std::future<int> result = task.get_future();
    std::thread task_td(std::move(task), 2, 10);
    task_td.join();
    std::cout << "task_thread:\t" << result.get() << '\n';
}

int main(){
    task_lambda();
    task_bind();
    task_thread();
    return 0;
}
```

最后我们用图 8.5-1 来展示异步可调用对象之间的关系。

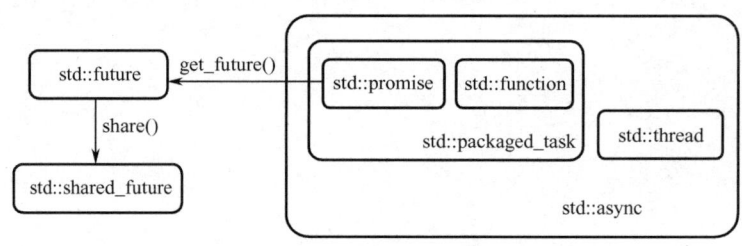

图 8.5-1　异步可调用对象之间的关系

8.6　线程池

线程池（thread pool）是一个编程概念，它本身不是 STL 的模板函数和类模板。线程池是一种在并发编程中经常使用的技术，用于管理和重用线程。线程池由线程池管理器、任务队列和线程池线程组成。

线程池的基本思想是，在应用程序启动时创建一组线程，并将它们保存在线程池的线程数组中。当需要执行任务时，从线程池中获取一个空闲的线程，将任务分配给该线程执行。当任务执行完毕后，线程将返回到线程池，随时被其他任务使用。

使用线程池的优点主要体现为以下几方面：

- 提高性能：线程池预先创建一定数量的线程并保存在内存中，这样可以避免频繁地创建和销毁线程，从而提高程序的性能。
- 资源管理：控制线程数量。通过线程池，可以控制同时运行的线程数量，避免系统资源的过度消耗。线程是系统级别的资源，如果线程数量过多，可能会导致系统资源的过度消耗，甚至可能导致系统崩溃。
- 任务调度：通过线程池，可以将任务分配给不同的线程执行，实现并行处理，从而使任务调度更加方便。

- 简化编程：使用线程池可以简化多线程编程的复杂性。开发人员只需要将任务提交给线程池，而不需要关心线程的创建、管理和销毁等细节，降低了多线程编程的难度。

下面的示例是一个简单线程池。

【例8-28】

```cpp
#include <condition_variable>
#include <iostream>
#include <mutex>
#include <queue>
#include <thread>
using namespace std;

class Task{
public:
    Task(){ arg_ = nullptr;}
    Task(std::string taskname):name_(taskname),arg_(nullptr){}
    Task(void *argument):arg_(argument){}
    void createTask(void (*handler)(void* arg), void*arg){
        this->handler = handler;
        this->arg_ = arg;
    }
    void (*handler)(void* arg);
    ~Task() {free(arg_);}
    void *arg_;
    std::string name_;
};

void worker(void *arg){
ThreadPool* pool = static_cast<ThreadPool*>(arg);

    while (true){
        //锁住队列以便数据可以被共享访问
        unique_lock<mutex> lock( pool->queue_mutex_);

        //当线程池停止并且任务队列没有任务时退出线程
        while(pool->tasks_.empty() || stop_){
            pool->cv_.wait(lock);
            return;
        }

        //从任务队列中取出下一个任务
        Task* task = move(pool->tasks_.front());
        Pool->tasks_.pop();
        lock.unlock();
        task-handler(task->arg_);         //执行任务
        delete task;
    }
}
```

```cpp
class ThreadPool{
public:
    ThreadPool(int num_threads){
        //创建工作线程,并把它们放入线程数组里
        nthreads_ = num_threads;
        for (size_t i = 0; i < nthreads_; ++i) {
            threads_[i] = std::thread(worker, this);
        }
    }
    ~ThreadPool(){
        {
            //锁住任务队列,更新 stop_标签
            unique_lock<mutex> lock(queue_mutex_);
            stop_ = true;
        }
        cv_.notify_all();                //通知所有线程

        //阻塞主线程,确保每一个工作线程都完成任务
        for (auto& thread : threads_) {
            if(thread.joinable())
                thread.join();
        }
    }

    //将任务放进任务队列中以便执行,并启用加锁机制
    void enqueue(Task* task){
        {
            unique_lock<std::mutex> lock(queue_mutex_);
            tasks_.emplace(move(task));
        }
        cv_.notify_one();
    }

    vector<thread> threads_;             //存放工作线程的向量
    queue<Task*> tasks_;                 //任务队列
    mutex queue_mutex_;                  //互斥量,同步对共享资源的访问
    condition_variable cv_;              //条件变量,用于表示任务队列状态的变化
    bool stop_ = false;                  //指示线程池是否停止的标志
    nthreads_;                           //线程池中的线程数
};

void factorial(void* size) {
    int* s = static_cast<int*>(size);
    int value = *s;
    long fact=1;
    for(int i=1; i<=value; i++) {
        fact*=i;
    }
    std::cout << "factoral_result = " << fact << std::endl;
```

```cpp
}

int main(){
    long * n = new int(5);
    ThreadPool pool(4);              //创建一个线程池,并初始化 4 个工作线程
    Task* t = new Task("Task1");
    t->createTask(factorial, n);
    pool.enqueue(t);

    *n = 10;
    t = new Task("Task2");
    t->createTask(factorial, n);
    pool.enqueue(t);

    *n = 15;
    t = new Task("Task3");
    t->createTask(factorial, n);
    pool.enqueue(t);

    *n = 20;
    t = new Task("Task4");
    t->createTask(factorial, n);
    pool.enqueue(t);

    delete n;
    return 0;
}
```

8.7 本章小结

本章比较系统地介绍了自 C++11 起 STL 中增加的有关并发与多线程的新内容,详细介绍了 C++11 的 STL 中有关多线程编程所涉及的模板类与模板函数,如 std::thread、std::mutex、std::condition_variable、std::atomic、std::lock_guard、std::unique_lock、std::lock、std::scoped_lock、std::async、std::future、std::promise、std::packaged_task 等,并给出了示例程序。

std::thread 是创建线程的类模板,它使用一个可调用对象(可以是普通函数、lambda 表达式、类成员函数以及函数对象等)来构造线程实例,并可以向可调用对象传入参数。

线程同步与互锁机制涉及的类与函数有 std::mutex、std::condition_variable、std::atomic、std::lock_guard、std::unique_lock、std::scoped_lock、std::lock。使用这些类与函数,能有效地实现多线程之间的通信、访问共享资源,从而使程序平稳运行而不会出现死锁问题。std::atomic 通过原子操作允许线程安全地访问和修改共享资源,而无须显式加锁或同步机制。

线程池是一种在多任务并发编程中经常使用的技术,用于管理和重用线程。线程池本身不是 STL 的函数模板和类模板,它由线程池管理器、任务队列和线程池线程组成。

std::async 是一个模板函数,也能够以异步或同步方式发起一个可调用对象。作为函数,std::async 的返回值是 std::future 对象,里面包含了可调用函数的结果或返回值。

std::future 和 std::promise 结合起来使用，可以实现两个线程之间的通信。std::promise 通常用于发送线程端，使用的成员函数是 set_value()。std::future 通常用于接收线程端，使用的成员函数是 get_it()。

std::packaged_task 和 std::function 类似，也可以绑定一个可调用对象，并使得被绑定的可调用对象被异步激活。std::packaged_task 的运行结果或者异常信息只能存储在一个共享状态中，只能通过 std::future 来访问。

第 9 章
并行算法与并行计算

9.1 STL 并行算法

为了从多核体系中受益，C++17 的标准模板库（STL）引入了并行算法来使用多个线程并行处理元素。STL 并行算法是 C++17 的最主要特色。C++17 引入了一些专门为并行编程补充的新算法。另外，许多算法都扩展了一个新的参数来指明是否要并行运行算法（当然，没有这个参数的旧版本仍然可使用）。

C++17 的并行算法定义了执行策略（execution policy）：顺序执行、并行执行和并行+向量化，并且提供了对应的执行策略类型和执行策略对象，如表 9.1-1 所示。

表 9.1-1　执行策略

执行策略类型（class）	std::execution::seq
	std::execution::par
	std::execution::par_unseq
	std::execution::unseq
执行策略对象（constant）	seqparpar_vec
动态执行策略（class）	execution_policy
测试一个类是否代表一个执行策略（class template）	is_execution_policy

9.1.1　std::execution::seq

std::execution::seq 和非并行化算法一样，当前线程会一个个地对所有元素执行操作。使用 std::execution::seq 和使用不接收执行策略参数的非并行化版本的效果类似，然而，std::execution::seq 比非并行化版本多了一些约束条件，如 for_each() 不能返回值、所有的迭代器必须至少是前向迭代器。C++17 提供 std::execution::seq 的目的是让程序员可以只修改一个参数来顺序地对元素进行操作，而不是使用一个不同签名的函数来顺序地对元素进行操作。注意：使用 std::execution::seq 的并行算法和相应的非并行化版本可能有些细微的不同。

9.1.2　std::execution::par

std::execution::par 意味着多个线程将会顺序地对元素执行操作。当某一线程对一个新元素进行操作前，该线程会先处理完它之前处理过的其他元素。与 std::execution::par_unseq 不同，

std::execution::par 可以保证在以下情况中不会出现问题或者死锁：执行了某个元素的第一步处理后，在执行另一个元素的第一步处理之前，必须执行这个元素后续的处理步骤。

9.1.3　std::execution::par_unseq

std::execution::par_unseq 意味着在执行多个线程时，不需要保证某个线程在执行完某一个元素的处理前不会被切换到其他的元素。特别地，std::execution::par_unseq 允许向量化执行，一个线程可以先执行多个元素的第一步处理，再执行下一步处理。std::execution::par_unseq 需要编译器/硬件的特殊支持来检测哪些操作如何向量化。

9.1.4　std::execution::unseq

std::execution::unseq 用于消除并行算法重载的歧义，并指示并行算法的执行可以向量化。例如，使用单个线程指令对多个数据进行操作。

大多数并行算法允许执行策略的重载。STL 的并行算法支持几种执行策略，STL 同时也提供了相应的执行策略类型和执行策略对象，用户可以静态地将所选的执行策略类型和执行策略对象作为参数来调用并行算法，也可以使用动态执行策略。

STL 支持的并行算法如图 9.1-1 所示，包括 69 个并行算法。

图 9.1-1　STL 支持的并行算法

实际上，STL 算法函数的并行策略是第一个参数是执行策略，例如：

```
std::vector <int> v = generate_large_vector();          //生成一个 vector
std::sort(v.begin(), v.end());                          //标准的升序排序，无并行策略
std::sort(std::parallel::seq, v.begin(), v.end());      //顺序执行
std::sort(std::parallel::par, v.begin(), v.end());      //容许并行执行
std::sort(std::parallel::par_unseq, v.begin(), v.end());//容许并行执行和向量化执行
```

9.2 常用的并行算法

本节主要介绍常用的并行算法，主要包括 std::sort、std::transform、std::find、std::find_if、std::find_if_not 和 std::search。注意，在使用 STL 的并行算法时，需要在程序中添加"#include <execution>"。

9.2.1 std::sort

std::sort 是 STL 算法库中比较常用的算法之一。STL 算法库中的算法都是在<algorithm>中定义的，这里只讨论 std::sort 的并行版本，其函数原型定义如下：

```
template<class ExecPolicy, class RandomIt, class Compare>
void sort(ExecPolicy&& policy, RandomIt first, RandomIt last, Compare comp);
```

其中：first、last 用来表示容器对象的元素范围；policy 是执行策略；comp 是二元比较函数对象，当第一个参数小于第二个参数时返回 true，如果没有指定 comp，则按照默认的升序进行排列。

【例 9-1】

```cpp
#include <algorithm>
#include <chrono>
#include <cstdint>
#include <iostream>
#include <random>
#include <vector>
#include <execution>

void measure(std::execution policy,std::vector<std::uint64_t> v){
    const auto start = std::chrono::steady_clock::now();
    std::sort(policy, v.begin(), v.end());
    const auto finish = std::chrono::steady_clock::now();
    std::cout<<std::chrono::duration_cast<std::chrono::milliseconds>(finish-start)<< '\n';
}

int main(){
    std::vector<std::uint64_t> v(1000000);
    std::mt19937 gen {std::random_device{}()};
    std::ranges::generate(v, gen);

    measure(std::execution::seq, v);
    measure(std::execution::unseq, v);
    measure(std::execution::par_unseq, v);
    measure(std::execution::par, v);
    return 0;
}
```

9.2.2　std::transform

std::transform 可将给定的操作函数应用到容器类对象的[first1, last1)内的每一个元素,并将结果保存到另一范围中,同时保持[first1, last1)内元素的顺序不变。std::transform 的非并行版本已经在第 6 章中介绍过了,这里只介绍其并行版本的原型定义。

```
template<class ExecPolicy, class ForwardIt1, Class ForwardIt2, class UnaryOperation>
ForwardIt2 transform(ExecPolicy&& policy, ForwardIt1 first1, ForwardIt1 last1, ForwardIt2 d_first,
                     UnaryOp unary_op);

template<class ExecPolicy,class ForwardIt1,class ForwardIt2, class ForwardIt3, class BinaryOperation>
ForwardIt3 transform(ExecPolicy&& policy, ForwardIt1 first1, ForwardIt1 last1, ForwardIt2 first2,
                     ForwardIt3 d_first, BinaryOp binary_op);
```

其中:first1、last1 用来表示容器类对象的元素范围;当操作对象为二元函数 binary_op 时,first2 表示第二组元素范围的起始位置;policy 表示执行策略;d_first 表示目标范围的起始位置,可以和 first1 或 first2 相同;unary_op 表示一元函数对象;binary_op 表示二元操作函数对象。

【例 9-2】

```cpp
#include <algorithm>
#include <iostream>
#include <array>
#include <vector>
#include <execution>

using namespace std;

struct Person {
    double height;
    double weight;
    int age;
    string_view name;
};

static constexpr array <Person, 8> people {
    { {1.72, 61.2, 23, "alex"},
    {1.88, 94.1, 26, "brad"},
    {1.65, 60.12, 22, "david"},
    {1.95, 110.3, 38, "derek"},
    {1.83, 79.5, 35, "jake"},
    {1.70, 64.2, 40, "jack"},
    {1.83, 96.3, 54, "sam"},
    {1.66, 80.5, 50, "tom"} }
};

struct PersonBmi{
    double bmi;
    string_view name;
```

```
};
int main() {
    vector<PersonBmi> vec_bmi;
    auto bmi = [](auto const& person){ return
    PersonBmi{person.weight / (person.height* person.height), person.name};};
    transform(std::execution::par, people.begin(), people.end(),
    back_inserter(vec_bmi),bmi);
    return 0;
}
```

【例 9-3】

```
#include <iostream>
#include <numeric>
#include <vector>
#include <execution>

int main() {
    std::vector v1 {1, 2, 3, 4, 5};
    std::vector v2 {10, 20, 30, 40, 50};
    std::vector v0;
    std::transform(std::execution::seq,
    v1.begin(), v1.end(),
    v2.begin(), back_inserter(v0), [](int i, int j) {return i*j;});
    return 0;
}
```

9.2.3　std::find、std::find_if 和 std::find_if_not

　　std::find 可在容器类对象的[first, last)内查找等于某个值的第一个元素，并在查到符合要求的元素后返回该元素的迭代器。std::find_if 可在容器类对象的[first, last)内查找满足条件 p 的第一个元素，并在查到第一个满足条件 p 的元素后返回该元素的迭代器。std::find_if_not 可在容器类对象的[first, last)内查找不满足条件 q 的第一个元素，并查找到第一个不满足条件 q 的元素后返回该元素的迭代器。三个查找函数的并行版本原型声明如下：

```
template<class ExecPolicy, class ForwardIt, class T>
ForwardIt find(ExecPolicy&& policy, ForwardIt first, ForwardIt last, const T& value);

template<class ExecPolicy, class ForwardIt, class UnaryOp>
ForwardIt find_if(ExecPolicy&& policy, ForwardIt first, ForwardIt last, UnaryOp p);

template<class ExecPolicy,class ForwardIt,class UnaryOp>
ForwardIt find_if_not(ExecPolicy&& policy, ForwardIt first, ForwardIt last, UnaryOp q);
```

　　其中：first、last 用来表示容器类对象的元素范围；policy 表示执行策略；p 表示一元函数对象，搜索到所需的元素时返回 true；q 表示一元函数对象，搜索到所需的元素时返回 false；value 是一个常量值，用于与容器内的元素逐一进行比较。

【例9-4】

```cpp
#include <algorithm>
#include <array>
#include <iostream>
#include <execution>

int main(){
    std::array<int, 5> v = {1, 2, 3, 4, 5};
    int n = 4;
    std::find(std::execution::seq, v.begin(), v.end(), n) == std::end(v))? std::cout <<"v does not contain "
            << n << endl;
    : std::cout << "v contains " << n << endl;
    auto is_even = [](int i){ return i % 2 == 0;};
    auto it = std::find_if(std::execution::seq, v.begin(), v.end(),is_even);
    if(it! = std::end(v))
    std::cout << "v contains an even number " << *it << endl;
    else
    std::cout << "v does not contain even numbers" << endl;
    std::array<int,5>::iterator iter;
    iter = std::find_if_not(std::execution::seq, v.begin(), v.end(), [](int i){return i%2;});
    std::cout << "The first even value is " << *iter << endl;
    return 0;
}
```

9.2.4　std::search

std::search 可在一个长序列里搜索子序列，长序列由 first、last 确定，子序列由 s_first、s_last 确定。std::search 有多个函数原型，支持并行计算的函数原型如下：

```cpp
template<class ExecPolicy,class ForwardIt1,class ForwardIt2>
ForwardIt1 search(ExecPolicy&& policy,ForwardIt1 first, ForwardIt1 last, ForwardIt2 s_first,
        ForwardIt2 s_last);

template<class ExecPolicy, class ForwardIt1, class ForwardIt2, class BinaryOp>
ForwardIt1 search(ExecPolicy&& policy, ForwardIt1 first, ForwardIt1 last, ForwardIt2 s_first,
        ForwardIt2 s_last, BinaryOp p);
```

其中：first、last 用来表示容器类对象元素的搜索范围，即长序列；s_first、s_last 用来表示容器类对象元素的搜索子序列；policy 表示执行策略；p 表示二元谓词函数，返回 bool 型的值。

【例9-5】

```cpp
#include <iostream>
#include <vector>
#include <algorithm>
#include <execution>

using namespace std;
int main(){
```

```cpp
    int i, j;
    vector<int> v1 = { 1, 2, 3, 4, 5, 6, 7 };
    vector<int> v2 = { 3, 4, 5 };
    vector<int>::iterator it;
    it = std::search(std::execution::seq, v1.begin(),v1.end(), v2.begin(),v2.end());
    if (it != v1.end()){
        cout <<"v2 is present at index: "<<(it-v1.begin())<< endl;
    }
    else {
        cout << "v2 is not present in v1" << endl;
    }
    return 0;
}
```

【例 9-6】

```cpp
#include <iostream>
#include <vector>
#include <algorithm>
#include <execution>
using namespace std;
bool pred(int i, int j){
    return (i > j)? 1 : 0;
}

int main(){
    int i, j;
    vector<int> v1 = { 1, 2, 3, 4, 5, 6, 7 };
    vector<int> v2 = { 3, 4, 5 };
    vector<int>::iterator it;
    it=std::search(std::execution::seq, v1.begin(),v1.end(), v2.begin(),v2.end(),pred);

    if (it != v1.end()) {
        cout<<"v1 elements are greater than v2 starting" << "from position " <<(it - v1.begin());
    }
    else {
        cout <<"v1 elements are not greater than v2 " << "elements consecutively.";
    }
    return 0;
}
```

9.3 C++17 中新增的并行算法

C++17 新增了一些并行算法函数（见表 9.3-1），这些并行算法支持并行策略，同时也保留了非并行版本。在使用并行算法的并行策略时，需要在程序中添加"#include <execution>"。本节将逐一介绍这些新增的并行算法。

表 9.3-1 C++17 新增的并行算法

并 行 算 法	所属的头文件	说　　　明
std::for_each	<algorithm>	std::for_each 出现异常时返回 void
std::for_each_n		应用函数对象于一个序列的前 n 元素
std::reduce（并行技术规范）	<numeric>	类似于 std::accumulate，出现异常时，次序变乱
std::exclusive_scan		类似于 std::partial_sum，计算前 i 个元素的和（不包括第 i 个元素）
std::inclusive_scan		类似于 std::partial_sum，计算前 i 个元素的和（包括第 i 个元素）
std::transform_reduce（并行技术规范）		应用一元函数对象先进行转换再进行求和
std::transform_exclusive_scan		应用一元函数对象然后计算和，排除输入
std::transform_inclusive_scan		应用一元函数对象，然后计算和，包括输入

9.3.1　std::for_each 和 std::for_each_n

std::for_each 是一个模板函数，有并行和非并行两种版本，将并行版本的第一个参数去掉即非并行版本。并行版本的原型定义如下：

```
template<class ExecPolicy,class ForwardIt,class UnaryFunc>
void for_each(ExecPolicy&& policy, ForwardIt first, ForwardIt last, UnaryFunc f);
```

其中：first、last 用于表示容器类对象元素的范围；f 表示函数对象，它应用到容器类对象的从 first 到 last 的每个元素；policy 表示 f 的执行策略。在 C++17 中，f 不能修改容器类对象元素的值。f 通常是 void 类型的，即没有返回值，如果 f 有返回值，则返回值会被忽略。

【例 9-7】

```cpp
#include <algorithm>
#include <iostream>
#include <vector>

int main(){
    std::vector<int> v {5, 10, 31, 20, 18, 92, 64};
    auto print = [](const int& i) { std::cout << i << ' '; };

    std::cout << "original: ";
    std::for_each(v.cbegin(), v.cend(), print);
    std::cout << endl;
    std::sort(v.begin(), v.end());
    std::cout << "after sorting: ";
    std::for_each(v.cbegin(), v.cend(), print);
    std::cout << endl;

    //对每个元素执行+=5 运算
    std::for_each(v.begin(), v.end(), [](int &i) {i+=5;});
    std::cout << "applying lambda to v: ";
    std::for_each(v.cbegin(), v.cend(), print);
```

```
            std::cout << endl;

            struct Sum{
                Sum(): sum(0){}
                void operator()(int n) { sum += n; }
                int sum;
            }s;

            //使用函数对象 Sum::operator()
            std::cout << "applying Sum::operator() to v: ";
            Sum ss = std::for_each(v.begin(), v.end(), s);
            std::cout << "sum: " << ss.sum << endl;
            return 0;
        }
```

在例 9-7 中，我们看到函数 f 有不同的形式。f 可以是 lambda 表达式，也可以是重载函数调用运算符 operator()的函数对象（这时 std::for_ach 的第三个参数将使用 f()形式，调用其重载函数调用运算符 operator()的函数对象）。

std::for_each_n 是 std::for_each 的另一种形式，即容器类对象元素范围是通过 first 加上 size 来确定的，二者仅此一点区别。

【例 9-8】

```
        #include <algorithm>
        #include <iostream>
        #include <vector>

        void print(auto const& v){
            for (auto const& e : v)
            std::cout << e << ", ";
        }

        int main(){
            std::vector<int> v {4, 1, 2, 6, 3, 8, 5, 9};
            print(v);
            std::cout << endl;

            std::for_each_n(v.begin(), 6, [](int& n) { n += 2;});
            std::sort(v.begin(), v.end());
            print(v);
            return 0;
        }
```

在例 9-8 使用的 lambda 表达式中，参数是通过引用传递的，在对 v 的前 6 个元素加 2 后，v 的前 6 个元素值将发生改变，随后按升序进行排序。

9.3.2　std::reduce 和 std::transform_reduce

std::reduce 是 C++17 引入的，有并行版本和非并行版本。非并行版本的 std::reduce 与

std::accumulate 相近。在并行版本中，std::reduce 根据 policy 对容器类对象的[first, last]内的元素进行归约。std::reduce 是在头文件<numeric>中定义的，并行版本函数原型定义有以下几种形式：

```
//第一种函数原型定义
template<class ExecPolicy, class ForwardIt>
typename std::iterator_traits<ForwardIt>::value_type
reduce(ExecPolicy&& policy, ForwardIt first, ForwardIt last);
//第二种函数原型定义
template<class ExecPolicy, class ForwardIt, class T>
T reduce(ExecPolicy&& policy, ForwardIt first, ForwardIt last, T init);
//第三种函数原型定义
template<class ExecPolicy, class ForwardIt, class T,
class BinaryOp>
T reduce(ExecPolicy&& policy, ForwardIt first, ForwardIt last, T init, BinaryOp binary_op);
```

其中：first、last 用于表示容器类对象的元素范围；policy 表示执行策略；init 表示累加器的初始值；binary_op 表示二元函数对象。

第一种函数原型定义等价于：

```
reduce(first,last,typename std::iterator_traits <InputIt>::value_type{});
```

第二种函数原型定义中使用了默认的二元函数对象 std::plus<>()，第三种函数原型定义中的 binary_op 采用的是指定的二元函数对象。

【例 9-9】

```cpp
#include <iostream>
#include <numeric>
#include <string>
#include <vector>
#include <execution>

int main(){
    std::vector v{32,16,8, 4, 2, 1};
    std::cout << std::reduce(std::execution::seq, v.begin(),
        v.end()) << endl;                                       //输出 63
    std::cout << std::reduce(std::execution::unseq, v.begin(),
        v.end(), 0) << endl;                                    //输出 63
    std::cout << std::reduce(std::execution::par, v.begin(),
        v.end(), 0, [](int a, int b){ return a+b;}) << enl;     //输出 63
    return 0;
}
```

在例 9-9 中，不同执行策略的 std::reduce 都得到相同的结果，但它们的运算过程是不同的。非并行版本的 std::reduce 与 std::accumulate 相似，std::accumulate 没有并行版本，它在做累加时是逐个将数组的元素和累加器相加的，并行版本的 std::reduce 则不同。

图 9.3-1 展示了并行版本的 std::reduce 和 std::accumulate 在进行累加时的区别，图中 std::reduce (std::execution::par)表示采用的并行策略是 std::execution::par。

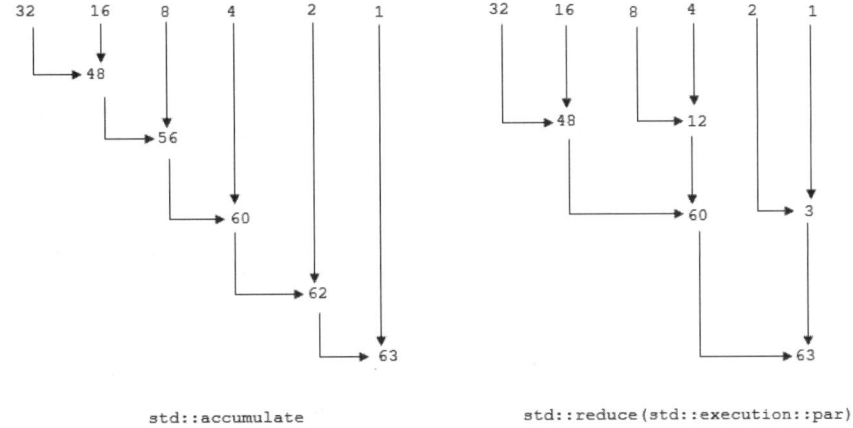

图 9.3-1　并行版本的 std::reduce 和 std::accumulate 在进行累加时的区别

【例 9-10】

```
#include <iostream>
#include <numeric>
#include <string>
#include <vector>
#include <execution>

int main(){
    std::vector v{64, 32, 16, 8, 4, 2, 1};

    std::cout << std::reduce(std::execution::seq, v.begin(),
    v.end()) << endl;        //输出 127
    std::cout << std::reduce(std::execution::unseq, v.begin()+1,
    v.end(), *v.begin(), std::minus<>()) << endl;
    //输出：53
    std::cout << std::reduce(std::execution::par, v.begin(),
    v.end(), 0, std::minus<>()) << enl;     //输出：23
    return 0;
}
```

std::transform_reduce 是 C++标准库<numeric>中的一个算法，它结合了转换（transform）和归约（reduce）两个操作。这个算法在处理容器元素时，先对元素进行转换操作，然后对转换后的结果进行归约操作，能高效地处理批量数据。std::transform_reduce 的并行版本函数原型如下：

```
//第一种函数原型定义
template< class ExecPolicy, class ForwardIt1, class ForwardIt2, class T>
T transform_reduce( ExecPolicy&& policy, ForwardIt1 first1, ForwardIt1 last1, ForwardIt2 first2, T init);
//第二种函数原型定义
template< class ExecPolicy, class ForwardIt1, class ForwardIt2, class T, class BinaryOp1, class BinaryOp2 >
T transform_reduce( ExecPolicy&& policy, ForwardIt1 first1, ForwardIt1 last1, ForwardIt2 first2, T init,
                    BinaryOp1 reduce, BinaryOp2 transform );
//第三种函数原型定义
template< class ExecPolicy, class ForwardIt, class T, class BinaryOp, class UnaryOp >
```

```
T transform_reduce(ExecPolicy&& policy, ForwardIt first, ForwardIt last, T init, BinaryOp reduce,
                   UnaryOp transform );
```

其中：first1、last1 用于表示容器 1 对象的元素范围；first2 表示容器 2 对象的起始元素；policy 表示执行策略；init 表示累加器的初始值；reduce 表示二元函数对象，以未指定的顺序应用到 transform 的返回结果；transform 表示将一元函数对象应用到容器对象的[first, last)内的每一个元素，并将返回结果输入 reduce。

在上面的第一种函数原型定义中，reduce 采用默认的 std::plus<>()，而 transform 采用默认的 std::multiply<>()。因此，第一种函数原型定义适合将两个容器对象的对应元素相乘，然后再相加，最终得到一个值。

【例 9-11】

```cpp
#include <vector>
#include <functional>
#include <iostream>
#include <numeric>
#include <execution>

int main()
{
    std::vector<int> v0(10000, 1);
    std::vector<int> v1(10000, 1);
    long result = std::transform_reduce(std::execution::par, v0.begin(), v0.end(),
                                        V1.begin(), 0);    //使用 std::plus<>()和 std::multiply<>()
    std::cout << result << '\n';           //输出 10000
    return 0;
}
```

【例 9-12】

```cpp
#include <iostream>
#include <numeric>
#include <vector>

int main() {
    std::vector v {1, 2, 3, 4, 5};
    std::cout << std::transform_reduce(v.begin(), v.end(), 0,
        [](int m, int n) {return m+n;},              //累加
        [](int i) {return i*i;})                     //转换
        << endl;                                     //输出 55
    std::cout << std::transform_reduce(v.begin(), v.end(), 0, [](int m, int n) {return m+n;},
            [](int i) {return i*i;}) << endl;        //先进行转换，再进行累加，最后输出 55

    return 0;
}
```

【例 9-13】

```cpp
#include <iostream>
#include <numeric>
```

```cpp
#include <vector>

int main() {
    std::vector v1 {1, 2, 3, 4, 5};
    std::vector v2 {10, 20, 30, 40, 50};
    std::cout << std::transform_reduce(v1.begin(), v1.end(), v2.begin(), 0, [](int m, int n) {return m+n;},
            [](int i, int j) {return i*j;}) << endl; //先进行转换，再进行累加，最后输出 550
    return 0;
}
```

9.3.3　std::exclusive_scan 和 std::inclusive_scan

std::exclusive_scan 和 std::inclusive_scan 都是 C++17 新增的算法，既有并行版本，也有非并行版本。std::exclusive_scan 和 std::inclusive_scan 类似于 std::partial_sum，可将容器类对象的[first, last)内的元素作为输入，输出范围的初始位置由 d_first 确定，求和的初始值由 init 确定。exclusive 的意思是输出结果将第 i 个元素排除在外，inclusive 的意思是输出结果将第 i 个元素包括在内。std::exclusive_scan 和 std::inclusive_scan 都是在<numeric>中定义的，都有多个函数原型，并行版本的函数原型定义如下：

```cpp
template<class ExecPolicy, class ForwardIt1, class ForwardIt2, class T, class BinaryOperation>
ForwardIt2 exclusive_scan(ExecPolicy&& policy, ForwardIt1 first, ForwardIt1 last, ForwardIt2 d_first,
        T init, BinaryOperation binary_op);
template<class ExecPolicy, class ForwardIt1, class ForwardIt2, class BinaryOperation, class T>
ForwardIt2 inclusive_scan(ExecPolicy&& policy, ForwardIt1 first, ForwardIt1 last, ForwardIt2 d_first,
        BinaryOperation binary_op, T init);
```

其中：first、last 用来表示容器类对象的元素范围；policy 表示执行策略；d_first 表示输出范围的起始位置，可以和 first 相同；init 表示累加器的初始值；binary_op 表示二元函数对象，以未指定的顺序应用到输入元素。

需要注意以下几点：

（1）在 std::exclusive_scan 中，init 也是目标序列的初始项；而在 std::inclusive_scan 中，init 只是初始值。

（2）当没有指定 binary_op 时，相当于使用 std::plus<>()对容器类对象的[first, last)内的元素进行累加，因而类似于 std::partial_sum。

（3）在 std::inclusive_scan 中，init 是可选项，因而放在最后一个参数。

【例 9-14】

```cpp
#include <functional>
#include <iostream>
#include <iterator>
#include <numeric>
#include <vector>
int main(){
    std::vector data {3, 1, 4, 1, 5, 9, 2, 6};

    std::cout << "Exclusive sum: ";
    std::exclusive_scan(std::execution::seq, data.begin(), data.end(),
```

```cpp
                    std::ostream_iterator<int>(std::cout, " "), 0);
    std::cout << "\nInclusive sum: ";
    std::inclusive_scan(std::execution::seq, data.begin(), data.end(),
                    std::ostream_iterator<int>(std::cout, " "));

    std::cout << "\n\nExclusive product: ";
    std::exclusive_scan(std::execution::seq, data.begin(), data.end(),
                    std::ostream_iterator<int>(std::cout, " "), 1, std::multiplies<>{});
    std::cout << "\nInclusive product: ";
    std::inclusive_scan(std::execution::seq, data.begin(), data.end(),
                    std::ostream_iterator<int>(std::cout, " "),
    std::multiplies<>{});
    return 0;
}
```

程序输出：

Exclusive sum: 0 3 4 8 9 14 23 25
Inclusive sum: 3 4 8 9 14 23 25 31

Exclusive product: 1 3 3 12 12 60 540 1080
Inclusive product: 3 3 12 12 60 540 1080 6480

【例 9-15】

```cpp
#include <functional>
#include <numeric>
#include <iostream>
#include <vector>

int main() {
    std::vector v{1, 2, 3, 4, 5, 6, 7, 8, 9, 10};
    std::vector<int> partial_sums{};
    partial_sums.reserve(v.size());
    std::vector<int> exclusion_scan_results{};
    exclusion_scan_results.reserve(v.size());
    std::vector<int> inclusive_scan_results{};
    inclusive_scan_results.reserve(v.size());

    std::partial_sum(v.begin(), v.end(),std::back_inserter(partial_sums));
    std::exclusive_scan(v.begin(), v.end(), std::back_inserter(exclusion_scan_results), 0, std::plus<int>());
    std::inclusive_scan(v.begin(), v.end(),std::back_inserter(inclusive_scan_results), std::plus<int>(), 0);

    std::cout << "partial_sum results    :";
    for (auto ps: partial_sums) {
        std::cout << ps << " ";
    }
    std::cout << std::endl;
    std::cout << "exclusive_scan results:";
    for (auto ps: exclusion_scan_results) {
        std::cout << ps << " ";
```

```cpp
    }
    std::cout << std::endl;
    std::cout << "inclusive_scan results:";
    for (auto ps: inclusive_scan_results) {
        std::cout << ps << " ";
    }
    std::cout << std::endl;
}
```

程序输出：

```
partial_sum results   :1 3 6 10 15 21 28 36 45 55
exclusive_scan results:0 1 3 6 10 15 21 28 36 45
inclusive_scan results:1 3 6 10 15 21 28 36 45 55
```

9.3.4　std::transform_exclusive_scan 和 std::transform_inclusive_scan

在理解 std::transform、std::exclusive_scan 和 std::inclusive_scan 后，读者就可以比较容易地理解 std::transform_exclusive_scan 和 std::transform_inclusive_scan。简单地讲，std::transform_exclusive_scan 和 std::transform_inclusive_scan 先对容器类对象的[first, last)内的元素进行转换，再将转换后的结果输入 std::exclusive_scan 或 std::inclusive_scan。std::transform_exclusive_scan 和 std::transform_inclusive_scan 都是在<numeric>中定义的，这两个并行算法带有执行策略，原型定义如下：

```cpp
template<class ExecPolicy,class ForwardIt1,class ForwardIt2, class T,class BinaryOp, class Unary Op>
ForwardIt2 transform_exclusive_scan(ExecPolicy&& policy, ForwardIt1 first, ForwardIt1 last,
                    ForwardIt2 d_first, T init, BinaryOp binary_op, UnaryOp unary_op);

template<class ExecPolicy,class ForwardIt1,class ForwardIt2, class BinaryOp, class UnaryOp, class T>
ForwardIt2 transform_inclusive_scan(ExecPolicy&& policy, ForwardIt1 first, ForwardIt1 last,
                    ForwardIt2 d_first, BinaryOp binary_op, UnaryOp unary_op, T init);
```

其中：first、last 用来表示容器类对象的元素范围；policy 表示执行策略；d_first 表示输出范围的起始位置，可以和 first 相同；init 表示累加器的初始值；unary_op 表示一元函数对象，作用在容器类对象指定范围内的每个元素；binary_op 表示二元函数对象，以未指定的顺序应用到输入元素。

需要注意几点：

（1）在 std::transform_exclusive_scan 中，init 也是目标序列的初始项；而在 std::transform_inclusive_scan 中，init 只是初始值。

（2）unary_op 先对容器类对象指定范围内的每个元素进行转换，再将转换后的结果输入 binary_op，binary_op 必须明确指定，不能省略。

（3）在 std::transform_inclusive_scan 中，init 是可选项，因而放在最后一个参数。

【例 9-16】

```cpp
#include <numeric>
#include <iostream>
#include <vector>
int main() {
```

```cpp
        std::vector v{1, 2, 3, 4, 5, 6, 7, 8, 9, 10};
        std::vector<int> results{};
        results.reserve(v.size());

        std::transform_exclusive_scan(v.begin(), v.end(), std::back_inserter(results), 0,
                                std::plus<int>{}, [](int i){return i*10;});

        for (auto r: results) {std::cout << r << " ";}
        std::cout << std::endl;

        results.clear();
        std::transform_inclusive_scan(v.begin(), v.end(), std::back_inserter(results),
                                std::plus<int>(), [](int i){return i*10;}, 0);

        for (auto r: results) {std::cout << r << " ";}
        std::cout << std::endl;
        return 0;
}
```

程序输出:

```
0 10 30 60 100 150 210 280 360 450
10 30 60 100 150 210 280 360 450 550
```

9.4 CUDA 并行计算编程

CUDA 是英伟达（NVIDIA）公司开发的并行计算平台和编程模型，用于在自己的 GPU（图形处理单元）上进行通用计算。CUDA 使开发人员能够利用 GPU 的强大功能来加速计算的可并行化部分。

本节通过一个简单的 CUDA 程序来介绍 CUDA 编程的几个基本概念，引导未接触过 CUDA 的读者入门 CUDA 编程。下面的示例把两个向量相加的程序转化为 CUDA 程序。

【例 9-17】

```cpp
#define N 10000000
void vector_add(float *out, float *a, float *b, int n){
    for(int i = 0; i < n; i++){
        out[i] = a[i] + b[i];
    }
}

int main(){
    float *a, *b, *out;
    a = (float*)malloc(sizeof(float)* N);
    b = (float*)malloc(sizeof(float)* N);
    out=(float*)malloc(sizeof(float)* N);

    for(int i=0; i< N; i++){
```

```
        a[i] = 1.0f;
        b[i] = 2.0f;
    }

    vector_add(out, a, b, N);
    return 0;
}
```

现在我们把上面的 C 语言程序改写成 CUDA 程序。

```
__global__ void vector_add(float *out, float *a, float *b,int n){
    for(int i = 0; i<n; i++){
        out[i] = a[i] + b[i];
    }
}

int main(){
    float *a, *b, *out;
    float *d_a, *d_b, *d_out;
    a = (float*)malloc(sizeof(float)* N);
    b = (float*)malloc(sizeof(float)* N);
    out = (float*)malloc(sizeof(float)* N);

    for(int i=0; i< N; i++){
        a[i] = 1.0f;
        b[i] = 2.0f;
    }

    cudaMalloc((void**)&d_a, sizeof(float) * N);
    cudaMalloc((void**)&d_b, sizeof(float) * N);
    cudaMalloc((void**)&d_out, sizeof(float) * N);

    cudaMemcpy(d_a, a, sizeof(float) * N, cudaMemcpyHostToDevice);
    cudaMemcpy(d_b, b, sizeof(float) * N, cudaMemcpyHostToDevice);

    vector_add<<<1,1>>>(d_out, d_a, d_b, N);
    cudaMemcpy(out, d_out, sizeof(float) * N, cudaMemcpyDeviceToHost);
    cudaFree(d_a);
    cudaFree(d_b);
    cudaFree(d_out);

    free(a);
    free(b);
    free(out);

    return 0;
}
```

上述 CUDA 程序的解释如下：
- device：设备，指 GPU 及其内存。

- host：主机，指 CPU 及其内存。
- __global__：函数前冠，__global__表明其后的函数是内核函数（kernel function）。内核函数在设备中运行，在主机中调用，如在 main()中调用。
- cudaMalloc、cudaMemcpy、cudaFree：分别表示在设备中分配内存、拷贝内存和释放内存。
- <<< M, N>>>：一对三重尖括号表明在调用内核函数时所使用的线程块数量，以及每个线程块中的线程数量。上述 CUDA 程序中的<<< 1, 1>>>表示使用了一个线程块和单个线程。

GPU 编程的大致流程如下：

（1）声明变量并在主机与设备上为变量分配内存。
（2）定义内核函数。
（3）在主机中初始化变量。
（4）将变量从主机传送至设备。
（5）执行一个或多个内核函数。
（6）将结果从设备传送回主机。

为了更好地理解 CUDA 程序，下面介绍几个概念。

- thread：线程。一个 CUDA 的并行程序会被分割成许多个线程来执行。
- block：线程块。多个线程会被组成一个线程块，同一个线程块中的线程可以同步，也可以通过共享内存（shared memory）进行通信。一个线程块中最多有 1024 个线程。线程在线程块中的索引用 threadIdx 表示，线程块中的线程以 3 个维度排列，即 threadIdx.x、threadIdx.y 和 threadIdx.z。
- grid：线程格，多个线程块可组成一个线程格。线程格中的线程块有 3 个维度，即 blockDim.x、blockDim.y 和 blockDim.z。每个维度的线程块有 3 个索引，即 blockIdx.x、blockIdx.y 和 blockIdx.z。同样线程格也有 3 个维度，即 gridDim.x、gridDim.y 和 gridDim.z。
- wrap：线程束。线程束是最基本的执行单元，一个线程束包含 32 个并行线程，这些线程在不同数据资源上执行相同的指令。

一个线程格的总线程数 N 为：

$$N = gridDim.x \times gridDim.y \times gridDim.z \times blockDim.x \times blockDim.y \times blockDim.z$$

通过 blockIdx.x、blockIdx.y、blockIdx.z 和 threadIdx.x、threadIdx.y、threadIdx.z 可以完全定位一个线程的位置了。当把所有的线程排成一个序列时，序列号为 0, 1, 2,…, N，如何才能找到当前的序列号呢？

（1）找到当前线程位于线程格中的哪一个线程块 blockId。

$$blockId = blockIdx.x + blockIdx.y \times gridDim.x + blockIdx.z \times gridDim.x \times gridDim.y$$

（2）找到当前线程位于线程块中的哪一个线程 threadId。

$$threadId = threadIdx.x + threadIdx.y \times blockDim.x + threadIdx.z \times blockDim.x \times blockDim.y$$

（3）计算一个线程块中的线程数量 M。

$$M = blockDim.x \times blockDim.y \times blockDim.z$$

（4）计算当前的线程序列号 idx。

$$idx = threadId + M \times blockId;$$

请注意：threadIdx、blockIdx、blockDim、gridDim 等是 CUDA 的保留字。

启动并行线程的步骤是：

（1）使用 kernel<<<N, M>>>（…）启动 N 个线程块，每个线程块有 M 个线程。
（2）使用 blockIdx.x 访问线程格内的线程块索引。

(3) 使用 threadIdx.x 访问块线程内的线程索引。

在一维数组的情况下,将元素分配给线程时,元素的索引为:

$$index = threadIdx.x + blockIdx.x \times blockDim.x$$

由上可知,两个向量相加的内核函数如下:

```
__global__ void add(int *a, int *b, int *c, int n){
    int index = threadIdx.x + blockIdx.x * blockDim.x;
    if (index < n)
        c[index] = a[index] + b[index];
}

#define N  512
__global__ void MatAdd(float A[N][N], float B[N][N], float C[N][N]){
    int i = blockIdx.x * blockDim.x + threadIdx.x;
    int j = blockIdx.y * blockDim.y + threadIdx.y;

    if (i < N && j < N)
    C[i][j] = A[i][j] + B[i][j];
}

int main(){
    //Kernel invocation
    dim3 threadsPerBlock(16, 16);              //指定每个线程块大小,dim3 为保留字
    dim3 numBlocks(N/threadsPerBlock.x, N/threadsPerBlock.y);
    MatAdd <<<numBlocks, threadsPerBlock>>>(A, B, C);
    ...
}
```

下面介绍共享内存与线程同步的概念。

每个线程块中都有共享内存(share memory),它的访问速度要比全局内存快很多,线程块内的所有线程都可以访问同一共享内存。__shared__ 表示其后的变量存储在共享内存里。

在线程之间共享数据时,需要注意避免竞争。虽然线程块中的线程在逻辑上是并行执行的,但并非所有的线程都可以同时在物理上执行。假设线程 A 和线程 B 分别从全局内存获取一个数据,并将该数据存储到共享内存中。线程 A 想要从共享内存中读取线程 B 的数据,反之亦然。假设线程 A 和线程 B 是两个不同线程束中的线程,如果线程 B 在线程 A 尝试读取它还没有写完的数据,就会产生一个竞争,这可能会导致未知的行为和不正确的结果。

为了确保并行线程协作时获得正确的结果,我们必须同步线程。CUDA 提供了一个简单的同步原语__syncthreads(),其功能是确保在某个线程块中的所有线程在执行到__syncthreads()之前都已完成它们之前的所有指令。只有所有的线程都到达__syncthreads(),它们才可以继续执行__syncthreads()之后的指令。__syncthreads()只能在设备代码(如内核函数)中使用,因此我们可以通过在存储到共享内存之后和从共享内存加载任何线程之前调用__syncthreads()来避免上述竞争。

在使用__syncthreads()时需要注意以下几点:

(1)在不同点调用__syncthreads()的结果是未知的,并且可能导致线程死锁,线程块中的所有线程必须在同一点调用__syncthreads()。

（2）不要在分支条件下不均匀地调用__syncthreads()，如果线程块中的一些线程调用了__syncthreads()，而其他线程由于某些条件（如 if 语句）没有调用__syncthreads()，那么可能会导致死锁。

（3）不要在循环中过度使用__syncthreads()，过度使用__syncthreads()可能会降低性能，因为__syncthreads()会阻止线程的并行执行。

【例9-18】

```
#include<stdio.h>

__global__ void staticReverse(int *d, int n){
    __shared__ int s[64];
    int t = threadIdx.x;
    int tr = n-t-1;
    s[t] = d[t];
    __syncthreads();
    d[t] = s[tr];
}

__global__ void dynamicReverse(int *d, int n){
    extern __shared__ int s[];              //extern 含义未变，表示 int s[]在外部定义.
    int t = threadIdx.x;
    int tr = n-t-1;
    s[t] = d[t];
    __syncthreads();
    d[t] = s[tr];
}

int main(void){
    const int n = 64;
    int a[n], r[n], d[n];

    for (int i = 0; i < n; i++) {
        a[i] = i;
        r[i] = n-i-1;
        d[i] = 0;
    }
    int *d_d;
    cudaMalloc(&d_d, n * sizeof(int));

    //run version with static shared memory
    cudaMemcpy(d_d, a, n*sizeof(int), cudaMemcpyHostToDevice);
    staticReverse<<<1,n>>>(d_d, n);
    cudaMemcpy(d, d_d, n*sizeof(int), cudaMemcpyDeviceToHost);

    for (int i = 0; i < n; i++)
    if(d[i]!= r[i])printf("Error:d[%d]!=r[%d](%d,%d)n",i,i,d[i], r[i]);

    //run dynamic shared memory version
```

```
    cudaMemcpy(d_d, a, n*sizeof(int), cudaMemcpyHostToDevice);
    dynamicReverse <<<1,n,n*sizeof(int)>>>(d_d, n);
    cudaMemcpy(d, d_d, n * sizeof(int), cudaMemcpyDeviceToHost);

    for (int i = 0; i < n; i++)
    if(d[i]!=r[i])printf("Error:d[%d]!=r[%d](%d,%d)n",i,i,d[i],r[i]);

    cudaFree(d_d);
    return 0;
}
```

GPU 可在矩阵运算方面发挥独特的优势，下面我们看看两个矩阵相乘的 CUDA 程序。

【例 9-19】

```
#include <curand.h>
#include <cublas_v2.h>
#include <iostream>

void print_matrix(const float *A, int rows, int cols){
    for(int i = 0; i < rows; ++i){
        for(int j = 0; j < cols; ++j){
            std::cout << A[j * rows + i] << " ";
        }
        std::cout << std::endl;
    }
    std::cout << std::endl;
}

void GPU_fill_rand(float *A, int nr_rows_A, int nr_cols_A) {
    //创建一个伪随机数生成器
    curandGenerator_t prng;
    curandCreateGenerator(&prng, CURAND_RNG_PSEUDO_DEFAULT);

    //使用系统时钟设置伪随机数生成器的种子
    curandSetPseudoRandomGeneratorSeed(prng, (unsigned long long) clock());

    //使用设备上的伪随机数填充数组
    curandGenerateUniform(prng, A, nr_rows_A * nr_cols_A);
}

void gpu_blas_mmul(const float *A, const float *B, float *C,
const int m, const int k, const int n) {
    int lda=m,ldb=k,ldc=m;
    const float alf = 1;
    const float bet = 0;
    const float *alpha = &alf;
    const float *beta = &bet;

    //Create a handle for CUBLAS ( CUDA 基本线性代数子程序库: CUDA Basic Linear Algebra Subroutine
```

library)

```
    cublasHandle_t handle;      //定义一个CUDA基本线性代数子程序库的句柄
    cublasCreate(&handle);      //创建一个CUDA基本线性代数子程序库的句柄

    //调用CUDA的cublas库的矩阵相乘函数
    cublasSgemm(handle, CUBLAS_OP_N, CUBLAS_OP_N, m, n, k, alpha, A, lda, B, ldb, beta, C, ldc);

    //销毁CUDA基本线性代数子程序库的句柄
    cublasDestroy(handle);
}

int main() {
    int A_rows, A_cols, B_rows, B_cols, C_rows, C_cols;
    A_rows = 4;
    A_cols = 3;
    B_rows = 3;
    B_cols = 4;
    C_rows = 4;
    C_cols = 4;

    //在主机内存中为三个矩阵分配内存
    float *h_A=(float *)malloc(A_rows * A_cols * sizeof(float));
    float *h_B=(float *)malloc(B_rows * B_cols * sizeof(float));
    float *h_C=(float *)malloc(C_rows * C_cols * sizeof(float));

    //在GPU中为三个矩阵分配内存
    float *d_A, *d_B, *d_C;
    cudaMalloc(&d_A, A_rows * A_cols * sizeof(float));
    cudaMalloc(&d_B, B_rows * B_cols * sizeof(float));
    cudaMalloc(&d_C, C_rows * C_cols * sizeof(float));

    //产生伪随机数矩阵A和矩阵B
    GPU_fill_rand(d_A, A_rows, A_cols);
    GPU_fill_rand(d_B, B_rows, B_cols);

    cudaMemcpy(h_A, d_A, A_rows * A_cols * sizeof(float),
    cudaMemcpyDeviceToHost);

    cudaMemcpy(h_B, d_B, B_rows * B_cols * sizeof(float),
    cudaMemcpyDeviceToHost);

    std::cout << "A =" << std::endl;
    print_matrix(h_A, nr_rows_A, nr_cols_A);
    std::cout << "B =" << std::endl;
    print_matrix(h_B, nr_rows_B, nr_cols_B);

    //在GPU上实现矩阵A和矩阵B的相乘
    gpu_blas_mmul(d_A, d_B, d_C, A_rows, A_cols, B_cols);
```

```
    //把数据从 GPU 的内存拷贝到主机内存
    cudaMemcpy(h_C, d_C, C_rows * C_cols * sizeof(float), cudaMemcpyDeviceToHost);

    std::cout << "C =" << std::endl;
    print_matrix(h_C, C_rows, C_cols);

    //Free GPU memory
    cudaFree(d_A);
    cudaFree(d_B);
    cudaFree(d_C);

    //Free CPU memory
    free(h_A);
    free(h_B);
    free(h_C);

    return 0;
}
```

在本节最后，我们介绍 CUDA 程序中的几个常见标识符。

- __global__：用于标识函数，表示该函数是内核函数，在设备中运行，在主机中调用。
- __device__：用于标识函数时，表示该函数在设备中运行，仅在设备中调用；用于标识变量时，表示该变量在设备中分配内存。
- __host__：用于标识函数，表示该函数在主机中运行，仅在主机中调用。
- __constant__：内存空间标识符，用于标识变量，表示定义的变量位于常量内存区间内，可与__device__一起用。
- __shared__：内存空间标识符，用于标识变量，表示定义的变量位于线程块的共享内存空间，可与__device__一起用。
- __managed__：内存空间标识符，用于标识变量，表示变量既可以在主机中引用，又可以在设备中引用，可与__device__一起用。
- __restrict__：用于标识限制性指针，限制性指针有助于编译器优化。
- __grid_constant__：用于标识内核函数参数，确保编译器不会在线程本地内存中创建内核参数的副本，而是使用参数本身的全局地址。

如果想要了解更详细的 CUDA 程序资料，可参考 NVIDIA 官网的 *CUDA Runtime API*、*CUDA C++ Programming Guide*。

9.5 OpenCL 编程

OpenCL（开放计算语言）是一种开放的、免版税的标准，用于超级计算机、云服务器、个人计算机、移动设备和嵌入式平台中的各种加速器进行跨平台的并行编程。OpenCL 极大地提高了各种应用程序的速度和响应能力，包括专业创意工具、科学和医疗软件、视觉处理，以及神经网络的训练和推理。

一个 OpenCL 应用程序在逻辑上可分为主应用程序（main application）和 OpenCL 程序（见图 9.5-1），主应用程序的代码使用通用编程语言（如 C 或 C++语言）编写，并由传统编译器编译。主应用程序在主机 CPU 上执行，包含调用主机 OpenCL API 来与 OpenCL 基础架构通信。OpenCL 基础架构为设备提供了通用的硬件抽象，包括硬件模型、内存模型和并行执行模型。设备无法运行传统 C 或 C++程序，OpenCL 基础架构提供的内核函数用于编写可以在设备上运行的程序。在设备上运行的 OpenCL 程序代码编译工作既可以在线完成，即在主应用程序执行期间使用特殊 API 来完成，也可以在执行主应用程序前将代码编译为机器的二进制码或由 Khronos 定义的特殊可移植中间件表示形式（称为 SPIR-V）。通过开源社区的努力，C++ for OpenCL 把 OpenCL 和 C++17 集成起来了。

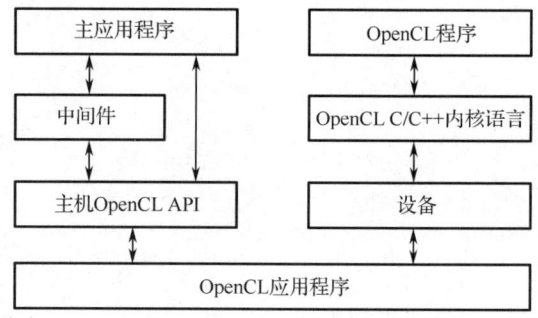

图 9.5-1　OpenCL 应用程序的逻辑组成

OpenCL 应用程序的主要实现步骤有：
（1）包含相关的头文件。
（2）检验所安装的 OpenCL 平台并设置一个设备。
（3）打开一个 OpenCL 上下文和一个应用程序代码。
（4）在 GPU 设备中创建缓存（分配内存）。
（5）创建一个命令队列（CommandQueue）并在其中放入设备命令。
（6）把数据写到设备。
（7）构造 OpenCL 的内核函数。
（8）在设备中调用该内核函数。
（9）从设备中取回数据。
（10）编译整个 OpenCL 应用程序代码。

【例 9-20】

```cpp
#include <iostream>
#include <CL/cl.hpp>                    //步骤（1）
int main(){
    //步骤（2）：获取所有的平台（驱动）
    std::vector<cl::Platform> all_platforms;
    cl::Platform::get(&all_platforms);

    if(all_platforms.size()==0){
        std::cout<<"Platforms is not installed.\n";
        exit(1);
    }
    cl::Platform default_platform = all_platforms[0];
```

```cpp
std::cout<<default_platform.getInfo<CL_PLATFORM_NAME>();

//获取默认平台的默认设备
std::vector<cl::Device> all_devices;
default_platform.getDevices(CL_DEVICE_TYPE_ALL, &all_devices);
if(all_devices.size()==0){
    std::cout<<"No devices found.Check OpenCL installation!\n";
    exit(1);
}
cl::Device default_device = all_devices[0];
std::cout<<"Device:" << default_device.getInfo <CL_DEVICE_NAME>();

//步骤（3）：打开一个 OpenCL 上下文（OpenCL Context）和应用程序代码（Program Source）
cl::Context context({default_device});
cl::Program::Sources sources;

//kernel calculates for each element C=A+B
std::string kernel_code = "void kernel simple_add(global const int* A, global const int* B,
                global int* C){ ""C[get_global_id(0)] = A[get_global_id(0)]+
                B[get_glbal_id(0)];""}";

sources.push_back({kernel_code.c_str(), kernel_code.length()});
cl::Program program(context, sources);

if(program.build({default_device})!=CL_SUCCESS){
    std::cout << program.getBuildInfo<CL_PROGRAM_BUILD_LOG>
    (default_device);
    exit(1);
}

//步骤（4）：在 GPU 设备中创建缓存
cl::Buffer buffer_A(context, CL_MEM_READ_WRITE, sizeof(int)*10);
cl::Buffer buffer_B(context, CL_MEM_READ_WRITE, sizeof(int)*10);
cl::Buffer buffer_C(context, CL_MEM_READ_WRITE, sizeof(int)*10);

int A[] = {0, 1, 2, 3, 4, 5, 6, 7, 8, 9};
int B[] = {0, 1, 2, 0, 1, 2, 0, 1, 2, 0};

//步骤（5）：创建 CommandQueue 并将其放入设备命令
cl::CommandQueue queue(context, default_device);

//步骤（6）：把数组 A 和数组 B 写入设备中
queue.enqueueWriteBuffer(buffer_A, CL_TRUE, 0, sizeof(int)*10, A);
queue.enqueueWriteBuffer(buffer_B, CL_TRUE, 0, sizeof(int)*10, B);

//步骤（7）：定义内核函数
cl::KernelFunctor simple_add(cl::Kernel(program, "simple_add"), queue, cl::NullRange,
                cl::NDRange(10), cl::NullRange);
```

```
                //步骤(8):调用内核函数
                simple_add(buffer_A, buffer_B, buffer_C);

                /*或者:步骤(7)和步骤(8)
                cl::Kernel kernel_add=cl::Kernel(program, "simple_add");
                kernel_add.setArg(0, buffer_A);
                kernel_add.setArg(1, buffer_B);
                kernel_add.setArg(2, buffer_C);

                queue.enqueueNDRangeKernel(kernel_add, cl::NullRange, cl::NDRange(10), cl::NullRange);
                queue.finish();
                */

                int C[10];
                //步骤(9):从设备读取数据到的数组C
                queue.enqueueReadBuffer(buffer_C, CL_TRUE, 0, sizeof(int)*10, C);

                std::cout<<" result: \n";
                for(int i=0;i<10;i++){ std::cout<<C[i]<<" ";}
                return 0;
            }
```

例 9-20 实现了数组 A 和数组 B 的相加,将结果放入数组 C。我们解释一下上面的程序。该示例选择了默认的平台 default_platform 和默认的设备 default_device。在设备中运行的内核函数是 simple_add(),该函数实现了两个数组的相加。main() 中的代码是在主机上运行的,simple_add 是在设备运行的。例 9-20 首先在设备上分配了缓存 buffer_A、buffer_B 和 buffer_C,代码如下:

```
cl::Buffer buffer_A(context, CL_MEM_READ_WRITE, sizeof(int)*10);
cl::Buffer buffer_B(context, CL_MEM_READ_WRITE, sizeof(int)*10);
cl::Buffer buffer_C(context, CL_MEM_READ_WRITE, sizeof(int)*10);
```

然后将 main() 中数组 A 和数组 B 的初始值复制到设备上的 buffer_A 和 buffer_B,代码如下:

```
queue.enqueueWriteBuffer(buffer_A,CL_TRUE,0,sizeof(int)*10,A);
queue.enqueueWriteBuffer(buffer_B,CL_TRUE,0,sizeof(int)*10,B);
```

接着将两个数组的相加结果从设备的 buffer_C 传回到数组 C 中,代码如下:

```
queue.enqueueReadBuffer(buffer_C, CL_TRUE, 0, sizeof(int)*10, C);
```

最后将数组 C 打印输出,代码如下:

```
for(int i=0;i<10;i++){ std::cout<<C[i]<<" ";}
```

再看将一组数相加的完整例子。

【例 9-21】

```
#define PROGRAM_FILE "add_numbers.cl"
#define KERNEL_FUNC "add_numbers"
#define ARRAY_SIZE 64

#include <math.h>
```

```c
#include <stdio.h>
#include <stdlib.h>
#include <string.h>
#include <time.h>

#ifdef MAC
#include <OpenCL/cl.h>
#else
#include <CL/cl.h>
#endif

cl_device_id create_device(){
    cl_platform_id platform;
    cl_device_id dev;
    int err;

    //识别一个平台
    err = clGetPlatformIDs(1, &platform, NULL);
    f(err < 0) {
        perror("Couldn't identify a platform");
        exit(1);
    }

    //访问一个设备
    err = clGetDeviceIDs(platform,CL_DEVICE_TYPE_GPU, 1, &dev, NULL);
    if(err == CL_DEVICE_NOT_FOUND){
        err=clGetDeviceIDs(platform, CL_DEVICE_TYPE_CPU, 1, &dev, NULL);
    }
    if(err < 0) {
        perror("Couldn't access any devices");
        exit(1);
    }
    return dev;
}

//打开一个文件,创建程序并编译它
cl_program build_program(cl_context ctx, cl_device_id dev, const char* filename){
    cl_program program;
    FILE *program_handle;
    char *program_buffer, *program_log;
    size_t program_size, log_size;
    int err;

    //读文件并且把内容放进缓存
    program_handle = fopen(filename, "r");
    if(program_handle == NULL){
        perror("Couldn't find the program file");
        exit(1);
    }
```

```c
        fseek(program_handle, 0, SEEK_END);
        program_size = ftell(program_handle);
        rewind(program_handle);
        program_buffer = (char*)malloc(program_size + 1);
        program_buffer[program_size] = '\0';
        fread(program_buffer, sizeof(char), program_size, program_handle);
        fclose(program_handle);

        //从文件创建程序
        program = clCreateProgramWithSource(ctx,1, (const char**)&program_buffer, &program_size, &err);
        if(err < 0) {
            perror("Couldn't create the program");
            exit(1);
        }
        free(program_buffer);

        //编译程序
        err = clBuildProgram(program, 0, NULL, NULL, NULL, NULL);
        if(err < 0) {
            //查找日志文件大小并打印输出
            clGetProgramBuildInfo(program, dev, CL_PROGRAM_BUILD_LOG, 0, NULL, &log_size);
            program_log = (char*) malloc(log_size + 1);
            program_log[log_size] = '\0';
            clGetProgramBuildInfo(program, dev, CL_PROGRAM_BUILD_LOG,
                                  log_size + 1, program_log, NULL);
            printf("%s\n", program_log);
            free(program_log);
            exit(1);
        }
        return program;
}

int main(){
    //OpenCL 应用程序结构
    cl_device_id device;
    cl_context context;
    cl_program program;
    cl_kernel kernel;
    cl_command_queue queue;
    cl_int i, j, err;
    size_t local_size, global_size;
    //数据与缓存
    float data[ARRAY_SIZE];
    float sum[2], total, actual_sum;
    cl_mem input_buffer, sum_buffer;
    cl_int num_groups;
    //初始化数据
    for(i=0; i<ARRAY_SIZE; i++) {
        data[i] = 1.0f*i;
```

```c
}
//创建设备与 OpenCL 上下文
device = create_device();
context = clCreateContext(NULL,1,&device,NULL,NULL,&err);
if(err < 0) {
    perror("Couldn't create a context");
    exit(1);
}

//编译程序
program = build_program(context, device, PROGRAM_FILE);
//创建数据缓存
global_size = 8;
local_size = 4;
num_groups = global_size/local_size;
input_buffer = clCreateBuffer(context, CL_MEM_READ_ONLY | CL_MEM_COPY_HOST_PTR,
                    ARRAY_SIZE*sizeof(float),data,&err);
sum_buffer = clCreateBuffer(context, CL_MEM_READ_WRITE | CL_MEM_COPY_HOST_PTR,
                    num_groups*sizeof(float),sum,&err);
if(err < 0) {
    perror("Couldn't create a buffer");
    exit(1);
};

//创建命令队列
queue = clCreateCommandQueue(context, device, 0, &err);
if(err < 0) {
    perror("Couldn't create a command queue");
    exit(1);
};

//创建一个内核函数
kernel = clCreateKernel(program, KERNEL_FUNC, &err);
if(err < 0) {
    perror("Couldn't create a kernel");
    exit(1);
};

//创建内核函数的参数
err =clSetKernelArg(kernel, 0, sizeof(cl_mem), &input_buffer);
err|=clSetKernelArg(kernel, 1, local_size*sizeof(float), NULL);
err|=clSetKernelArg(kernel, 2, sizeof(cl_mem), &sum_buffer);
if(err < 0) {
    perror("Couldn't create a kernel argument");
    exit(1);
}

//将内核函数加入命令队列
Err = clEnqueueNDRangeKernel(queue, kernel, 1, NULL, &global_size, &local_size, 0,
```

```
                        NULL, NULL);
    if(err < 0){
        perror("Couldn't enqueue the kernel");
        exit(1);
    }

    //读内核函数的输出
    err = clEnqueueReadBuffer(queue, sum_buffer, CL_TRUE, 0, sizeof(sum), sum, 0, NULL, NULL);
    if(err < 0) {
        perror("Couldn't read the buffer");
        exit(1);
    }

    //检查结果
    total = 0.0f;
    for(j=0; j<num_groups; j++) {
        total += sum[j];
    }
    actual_sum = 1.0f * ARRAY_SIZE/2*(ARRAY_SIZE-1);
    printf("Computed sum = %.1f.\n", total);
    if(fabs(total - actual_sum) > 0.01*fabs(actual_sum))
    printf("Check failed.\n");
    else
    printf("Check passed.\n");

    //释放内存资源
    clReleaseKernel(kernel);
    clReleaseMemObject(sum_buffer);
    clReleaseMemObject(input_buffer);
    clReleaseCommandQueue(queue);
    clReleaseProgram(program);
    clReleaseContext(context);
    return 0;
}
```

9.6 本章小结

本章首先介绍了 C++17 引入的并行算法的基本概念和几种执行策略，如 std::parallel::seq、std::parallel::par、std::parallel::par_unse 和 std::parallel::unseq。

从 C++17 起，STL 算法库中有很多算法都增加了执行策略，本章在 9.2 节中介绍了几种常用算法的执行策略（如 std::sort、std::transform、std::find 和 std::search）的应用。

本章在 9.3 节中通过示例介绍了 STL 算法库中新增的几种并行算法（如 std::for_each、std::for_each_n、std::reduce、std::transform_reduce、inclusive_scan、std::exclusive_scan、transform_exclusive_scan、transform_inclusive_scan）的应用。这些并行算法在形式上都是模板函数，而不是模板类。这些并行算法的作用对象基本上是 STL 的容器类对象，特别是 std::vector，因而有着相

似的函数原型，并通常将执行策略作为第一参数。去掉执行策略，并行化版本的函数原型就变成了非并行化版本的函数原型。

CUDA 是英伟达公司推出的目前在工业界很受欢迎的基于 GPU 的并行计算框架。本章在 9.4 节中介绍了 CUDA 并行计算编程的基本概念、步骤与方法，并通过示例介绍了 CUDA 程序的一些重要的概念，如线程、线程块、线程格、线程索引，以及共享内存和线程同步，帮助未接触过 CUDA 的读者快速入门 CUDA 编程。

OpenCL 是一种开放的、免版税的、跨平台的并行计算框架，目前正在工业界备受欢迎。本章在 9.5 节中简单介绍了 OpenCL 编程的基本步骤与方法。

CUDA 并行计算编程与 OpenCL 编程非常相似，在实际编程中，CPU 及内存构成了主机，GPU 及内存构成了设备。内核函数是并行计算的核心，它是在设备上存储和运行的。也就是说，要计算的数组或者矩阵都需要在设备上分配存储空间，并进行并行计算。

OpenCL 的并行计算具有明显的优势，主要是因为通过开源社区的努力，C++ for OpenCL 把 OpenCL 和 C++17 集成起来了，从而使面向对象的并行计算编程更符合习惯。

本章是从编程的角度介绍并行算法和并行计算的，没有涉及并行算法和并行计算的理论和方法。

第 10 章 设计模式

10.1 设计模式概念

C++中的设计模式可帮助开发人员创建可维护、灵活且易于理解的代码。设计模式囊括了软件架构师和开发人员的专业知识及丰富经验，可以使新程序员更容易地遵循既定的最佳实践。设计模式是软件工程中针对经常出现的问题而设计的可通用、可重复的解决方案，它不是一个可以立即用代码编写的完整设计，而是一种解决问题的描述或模型，可以应用于各种设计中。

设计模式通常有三种，即创建设计模式（creational design pattern）、结构设计模式（structural design pattern）和行为设计模式（behavioral design pattern）。这三种设计模式覆盖了软件设计的主要方面，并且每一种设计模式又可细分为多种设计方法。本章将对这三种设计模式中的每一种设计方法进行详细的介绍。针对每一种设计方法，本章都给出问题描述、解决方案、UML 类图，以及相应的 C++实例代码。

本章所介绍的每一种设计模式都会使用 UML 类图（class diagram）进行表达，因此在正式介绍设计模式之前，先简要介绍一下 UML 类图。图 10.1-1 给出了一个简单的 UML 类图示例。

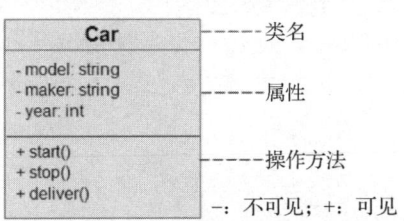

图 10.1-1 UML 类图示例

在 UML 类图中，一个类由类名、属性（attribute）和操作（operation）/方法（method）三部分组成。属性和操作方法有可见（用+表示）与不可见（用-表示）之分。在 C++中，可见对应 public，不可见对应 private 和 protected。

在 UML 类图中，类之间的关系有泛化（generalization）、聚合（aggregation）、有向关联（directed association）、组合（composition）、关联（association）、实现（realization）、依赖（dependency）和注释连接器（note connector）。UML 类图中类之间关系的标识符如图 10.1-2 所示。

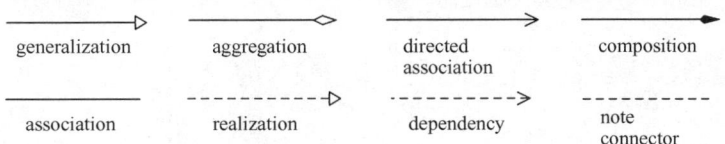

图 10.1-2 UML 类图中的各种关系标识符

在本章介绍的设计模式中，泛化和实现是最常用的两种关系。图 10.1-3（a）所示为泛化关系，我们可以从 Checking、Saving 和 Credit 三种具体的银行账户泛化出一般性的 Account。在 C++中，Account 一般被声明为基类，里面存放公共的属性与方法。Checking、Saving 和 Credit 均为 Account 的派生类。每个派生类继承了基类的属性与方法，并且定义了自己特有的属性和方法。图 10.1-3（b）所示为实现关系，图中<<interface>>是接口类。C++没有接口关键字。什么是接口类呢？接口类是指类中至少有一个函数是纯虚函数的类，例如：

virtual void fly() = 0;

注意：接口类是不能被实例化的，因此接口类中的所有函数都需要声明为纯虚函数。从接口类派生出的具体类（如 Bird）需要实现接口类中所有声明的接口函数。

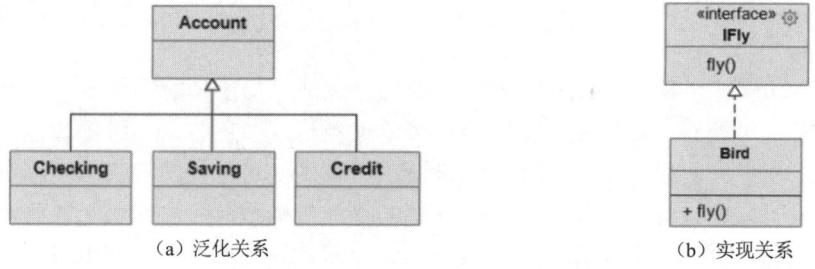

图 10.1-3　泛化关系与实现关系

图 10.1-4（a）所示为聚合关系，图中的类 Car 与类 Engine、类 Wheel 是聚合关系。聚合关系不是拥有关系，也不是父子关系，表示一个类（如 Car）是另一个类（如 Engine）的非排他性容器。在 C++中，聚合关系通常表现为整体与部分的关系，部分是整体的成员变量。图 10.1-4（b）所示为组合关系，图中的类 Person 与类 Hand、类 Head、类 Leg 是组合关系。组合关系是一种强聚合关系，强调整体与部分的依赖关系和不可分割性。在 C++中，类 Hand、类 Head、类 Leg 也是类 Person 的成员变量。在图 10.1-4 中，类 Car 和类 Person 不是基类，更不是接口类，类 Engine 和类 Wheel 不是从类 Car 派生出来的，类 Hand、类 Head 和类 Leg 也不是从类 Person 派生出来的。

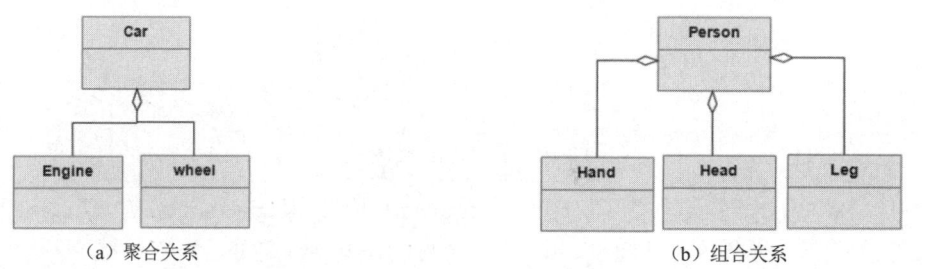

图 10.1-4　聚合关系和组合关系

10.2　创建设计模式

创建设计模式是设计模式的一个子集，它涉及对象的创建方法与创建过程，试图使对象创建变得更加灵活和高效。创建设计模式主要包括工厂方法（factory method）、抽象工厂方法（abstract factory method）、构建器方法（builder method）、原型方法（prototype method）和单例方法（singleton method）。

10.2.1 工厂方法

工厂方法是一种创建设计模式，它定义了一个创建对象的接口类，并由该接口类的派生类来创建对象的实例。

假设我们要开发一个物流系统，该物流系统有陆地物流（主要是卡车运输）和海运物流（主要是轮船海运）和空中物流（主要是飞机运输）三个部分，为此我们设计三个类，即 Truck、Ship 和 Airplane，这三个类负责具体的物流运输。我们还需要设计一个接口类 Transport，它是类 Truck、类 Ship 和类 Airplane 的基类。注意：接口类在 C++中被定义为抽象类。该接口类声明了一个名为 deliver()的接口函数。在 C++中，接口函数一般被定义为纯虚函数。类 Truck、类 Ship 和类 Airplane 以不同的方式实现 deliver()。例如，在实际的物流系统中，卡车用箱体通过陆路运送货物，船舶通过集装箱海上运送货物，飞机用一般箱体快速运送货物。工厂方法的 UML 类图示例如图 10.2-1 所示。

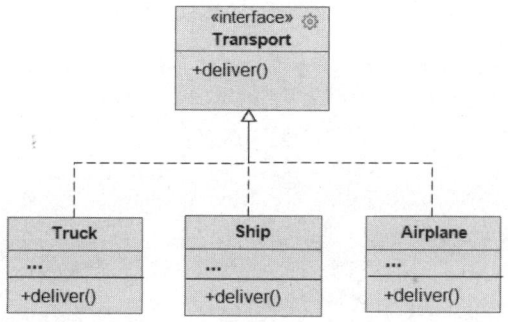

图 10.2-1　工厂方法的 UML 类图示例

在使用工厂方法创建类 Truck、类 Ship 和类 Airplane 的实例时，不是在 Client 直接使用 new 运算符创建实例的，而是在特殊工厂方法的内部创建实例的，虽然也是通过 new 运算符来创建的，但特殊工厂方法是从工厂方法中调用的。我们先定义一个特殊的工厂基类 TransportCreator，然后从它派生三个类，即 TruckCreator、ShipCreator 和 AirplaneCreator，最后由三个派生类负责创建类 Truck、类 Ship 和类 Airplane 的实例。例 10-1 给出了物流系统创建对象的示范代码。

【例 10-1】

```
#include<iostream>
#include<string>

class Transport{
public:
    virtual ~Transport () {}
    virtual std::string delivery() const = 0;            //纯虚函数
};

class Truck : public Transport{
public:
    std::string delivery() const override {
        return "{Delivery by trucks}";
    }
```

```cpp
};

class Ship: public Transport{
public:
    std::string delivery() const override{
        return "{Delivery by ships}";
    }
};

class Airplane : public Transport{
public:
    std::string delivery() const override {
        return "{Delivery by airplanes}";
    }
}

class TransportCreator{
public:
    virtual ~TransportCreator() {delete p_transport;}
    virtual Transport* TransportFactory() const = 0;          //纯虚函数
    void TransportDelivery(){
        //通过纯虚函数创建具体的对象，并调用 delivery()
        p_transport = TransportFactory();
        p_transport->delivery();
    }
private:
    Transport* p_transport;
};

class TruckCreator : public TransportCreator{
public:
    transport* TransportFactory() const override{
        return new truck();
    }
};

class ShipCreator : public TransportCreator{
public:
    transport* TransportFactory() const override{
        return new ship();
    }
};

class AirplaneCreator : public TransportCreator{
public:
    transport* TransportFactory() const override{
        return new airplane();
    }
};
```

```cpp
void ClientCode(std::shared_ptr<TransportCreator> pCreator){
    std::cout << pCreator->TransportDelivery() << std::endl;
}

int main(){
    std::shared_ptr<TransportCreator> truckcreator(new TruckCreator());
    ClientCode(truckcreator);
    std::cout << std::endl;
    std::shared_ptr<TransportCreator> shipcreator(new ShipCreator());
    ClientCode(shipcreator);
    std::cout << std::endl;
    std::shared_ptr<TransportCreator> airplanecreator(new AirplaneCreator());
    ClientCode(airplanecreator );

    return 0;
}
```

10.2.2 抽象工厂方法

抽象工厂方法是一种创建设计模式，它提供了一个接口，用于创建一组相关或依赖的对象，而无须指定其具体类。

假设我们需要为一家 OfficeWorks（购买办公用品的大型连锁商店）开发一款 App。
- 产品种类：Chair（椅子）、Desk（办公桌）、File Cabinet（文件柜）等。
- 产品样式：Chinese Style（中国样式）和 European Style（欧洲样式）。

现在我们使用抽象工厂方法来创建具体产品的对象。首先我们需要为每一种产品定义一个接口类。在 C++语言中，我们先定义本例中的三个接口类，即 Desk、Chair 和 Cabinet。而类 ChineseStyleDesk、类 EuropeanStyleDesk 是派生类。FurnitureFactory 是一个接口类，类中只有接口函数。我们声明了类 ChineseFurnitureFactory 和类 EuropeanFurnitureFactory，它们要实现 FurnitureFactory 中的接口函数。通过虚函数机制会自动创建对应样式产品的实例。抽象工厂方法的 UML 类图示例如图 10.2-2 所示，例 10-2 通过具体的代码来展示抽象工厂方法的应用。

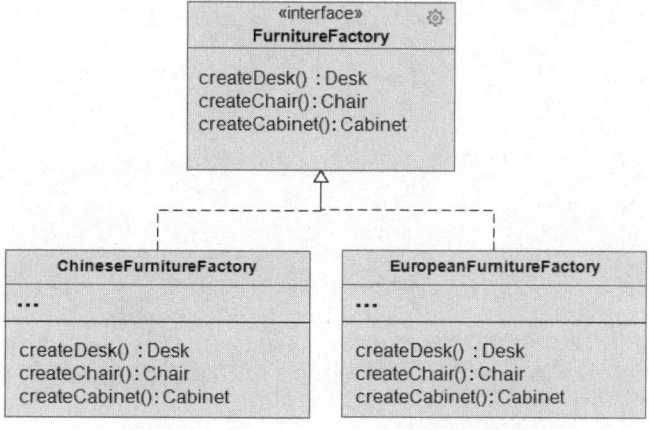

图 10.2-2 抽象工厂方法的 UML 类图示例

【例 10-2】

```cpp
#include<iostream>
#include<string>

class Desk{
public:
virtual ~Desk(){};
    virtual std::string DeskInformation() const = 0;
};

class ChineseStyleDesk : public Desk{
public:
std::string DeskInformation() const override{
        return "The information of ChineseSyleDesk.";
    }
};

class EuropeanStyleDesk : public Desk{
public:
std::string DeskInformation() const override {
        return "The information of EuropeanStyleDesk.";
   }
};

class Chair{
public:
    virtual ~Chair(){};
    virtual std::string ChairInformation() const = 0;
};

class ChineseStyleChair : public Chair{
public:
    std::string ChairInformation() const override {
        return "The information of ChineseStyleChair";
    }
};

class EuropeanStyleChair : public Chair{
public:
    std::string ChairInformation() const override {
        return "The information of EuropeanStyle.";
}
};

//接口类
class FurnitureFactory {
public:
virtual Desk* CreateDesk() const = 0; //纯虚函数
```

```cpp
    virtual Chair* CreateChair() const = 0; //纯虚函数
};

class ChineseFurnitureFactory : public FurnitureFactory{
public:
    Desk* CreateDesk() const override{
        return new ChineseStyleDesk();
    }
    Chair *CreateChair() const override{
        return new ChineseStyleChair();
    }
};

class EuropeanFurnitureFactory : public FurnitureFactory{
public:
Desk *CreateDesk() const override{
        return new EuropeanStyleDesk();
}
Chair *CreateChair() const override{
        return new EuropeanStyleChair();
    }
};

void ClientCode(FurnitureFactory* factory){
    Desk *pDesk = factory->CreateDesk();
    Chair *pChair = factory->CreateChair();
    std::cout << pDesk->DeskInformation() << "\n";
    std::cout << pChair->ChairInformation() << "\n";
    delete pDesk;
    delete pChair;
}

int main(){
    std::cout << "Create Chinese style furniture\n";
    FurnitureFactory * f1 = new ChineseFurnitureFactory();
    ClientCode(f1);
    std::cout << std::endl;
    std::cout << "Create European style furniture:\n";
    FurnitureFactory* f2 =   new EuropeanFurnitureFactory();
    ClientCode(f2);
    delete f1;
    delete f2;
    return 0;
}
```

10.2.3　构建器方法

构建器方法是一种用于创建复杂对象的创建设计模式，它提供了一个接口类，然后由具有实现该接口类的具体构建器（ConcreteBuilder）类来创建特定的对象。

本节以创建复杂对象为例给出构建器方法的 UML 类图和相应的代码示例。例如，我们要创建一个对象，它包括发动机系统、传动系统、电路系统、冷却系统、装载系统等组成部分。我们不想为类 Tractor 定义一个巨大的构造函数，将几百种零部件参数放入其中。我们可以采用一个简单的解决方案：扩展类 Tractor 并创建一组部件类，如类 Engine、类 Transmission 等。构建器方法通常包含以下部分：

（1）Product：是所需创建的复杂对象，有较多的属性和组件。

（2）Builder：是一个接口类，定义了创建复杂对象的接口函数。在 C++中，接口类是一个抽象类，接口函数是纯虚函数。

（3）ConcreteBuilder：是一个具体的构建器，在 C++中，ConcreteBuilder 是类 Builder 的派生类，实现了类 Builder 中定义的接口细节。ConcreteBuilder 可以有多个。

（4）Director：负责管理创建复杂对象的过程，它不关心也不必知道复杂对象的每一个组件细节，只负责在上层管理接口来构建复杂对象。

（5）Client：发起创建复杂对象的函数。

构建器方法的 UML 类图示例如图 10.2-3 所示。

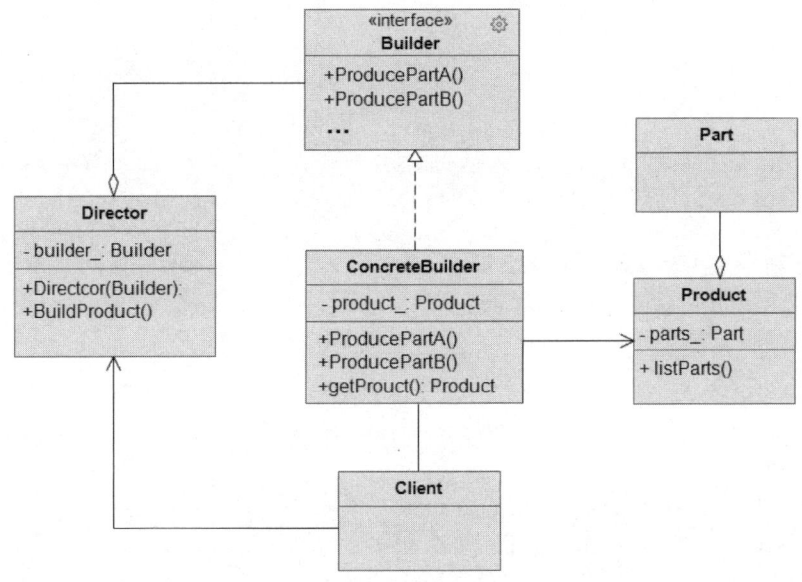

图 10.2-3　构建器方法的 UML 类图示例

【例 10-3】

```
#include<iostream>
#include<vector>
#include<string>

class Part{
public:
    Part(const std::string part_name_){part_name_ = part_name;}
    virtual void getPartInfo() { cout << part_name_ ;}
    protected:
```

```cpp
    std::string part_name_;
};

class Engine: public Part{
public:
    Engine(std::string producer):Part("engine"){ producer_ = producer ;}
    void getPartInfo(){std::cout << part_name_ <<","<<producer_ << endl;}
private:
    std::string producer_;
};

class Transmission: public Part{
public:
    Transmission(std::string producer) : Part("transmission"){
        producer_ = producer;
    }
    void getPartInfo(){std::cout << part_name_<<","<<producer_<< endl;}
private:
    std::string producer_;
};

class Tractor{
public:
    std::vector<Part*> tractor_parts_;
    void listParts() const{
        std::cout << "Tractor parts: ";
        for (size_t i = 0; i < tractor_parts_.size(); i++){
            if(tractor_parts_[i] == tractor_parts_.back()){
                std::cout << tractor_parts_[i]->getPartInfo();
            }else{
                std::cout << tractor_parts_[i] ->getPartInfo() << ", ";
            }
        }
        std::cout << "\n\n";
    }
};

class Builder{
public:
    virtual ~Builder(){}
    virtual void ProducePartA(const std::string sPartA) const =0;
    virtual void ProducePartB(const std::string sPartB) const =0;
};

class TractorBuilder : public Builder{
private:
    Tractor* tractor_;
public:
    TractorBuilder(){
```

```cpp
            tractor_ = new Tractor();
        }
        ~TractorBuilder(){
            delete tractor_;
        }
        void ProducePartA(const std::string sProducer) const final{
            Part *p   = new Engine(sProducer);
            tractor_->tractor_parts_.push_back(p);
        }

        void ProducePartB(const std::string sProducer)const final{
            Part *p   = new Transmission(sProducer);
            tractor_->tractor_parts_.push_back(p);
        }

        Tractor* GetProduct() {
            return tractor_;
        }
};

Class Director{
private:
    Builder* builder;
    std::string engine_producer;
    std::string transmission_producer;
public:
    void set_builder(Builder* builder){
        this->builder=builder;
    }
    void setEngineProducer(const std::string producer) {
        engine_producer = producer;
    }
    void setTransmissionProducer(const std::string producer){
        transmission_producer = producer;
    }
    void BuildTractor(){
        this->builder->ProducePartA(engine_producer);
        this->builder->ProducePartB(transmission_producer);
    }
};

void ClientCode(Director& director){
    TractorBuilder* builder = new TractorBuilder();
    director.set_builder(builder);
    std::cout << "Build product tractor:\n";
    director.BuildTractor();
    Tractor* p= builder->GetProduct();
    p->ListParts();
    delete builder;
```

```
}
int main(){
    Director* director= new Director();
    director->setEngineProducer("GE");
    director->setTransmissionProducer("Toyota");
    ClientCode(*director);
    delete director;
    return 0;
}
```

10.2.4 原型方法

原型方法是一种创建设计模式，用于创建与现有对象具有相同结构和初始状态的新对象，从而创建了现有对象的精确副本，而无须显式地指定构造细节。

假设我们已经有了一个对象，但现在想要创建它的精确副本。首先，我们放弃传统的做法：创建该类的新对象，然后遍历原对象的所有属性，并将属性的值复制到新对象。其次，在现实中并非对象的所有属性都可以用这种方式复制，某些属性可能是私有的，从对象的外部来看是不可见的。

原型方法采用克隆方法，支持克隆的对象称为原型对象。原型方法为所有支持克隆的对象声明一个通用接口，此接口供 Client 调用。通常，这个通用接口仅包含单个克隆方法。克隆方法的实现在所有类中都非常相似，这是因为在 C++中，将*this 代入拷贝构造函数时，便实现了克隆。克隆是一种很有效的对象创建方法，特别是当原型对象具有数十个字段时。

原型方法的 UML 类图示例如图 10.2-4 所示。在 C++中，类 Vehicle 是纯抽象基类，只定义了一个接口函数 Clone()（该接口函数是纯虚函数）。类 Car 和类 Truck 是实际克隆的对象，它们的 Clone()负责具体的克隆工作。这里类 VehicleFactory 是包装了类 Car 和类 Truck 的克隆工厂。

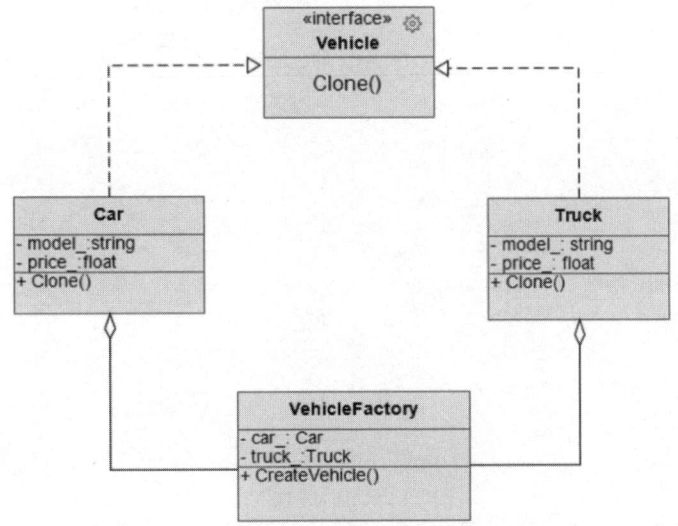

图 10.2-4　原型方法的 UML 类图示例

在我们给出的示例代码中，类 Vehicle 是纯抽象基类，它只定义了接口函数 Clone()，但无法实现。类 Car 和类 Truck 是从纯抽象基类 Vehicle 派生而来的，它们的 Clone()负责具体的克隆工作，创建精确副本。类 VehicleFactory 是克隆工厂，它先创建了两个用于克隆的原型对象；然后在克隆对象时把原型对象传入，由两个原型对象克隆自己。

【例 10-4】

```cpp
#include<stream>
#include<string>
using namespace std;

enum class Type {
    VEHICLE_1 = 0,
    VEHICLE_2
};

class Vehicle {
public:
    Vehicle() {}
    virtual ~Vehicle() {}
    virtual Vehicle *Clone() const = 0;
};

class Car : public Vehicle{
private:
    string model_;
    float price_;
public:
    Car(string model_name, float price) : model_(model_name), price_(price){}

    Vehicle *Clone() const override {
        return new Car(*this);                    //克隆一个 Car
    }
};

class Truck : public Vehicle{
private:
    string model_;
    float price_;

public:
    Truck(string model_name, float price) : model_(model_name), price_(price){}

    Vehicle *Clone() const override{
        return new Truck(*this);                  //克隆一个 Truck
    }
};

class VehicleFactory{
private:
    map<Type, Vehicle *> vehicles_;
public:
```

```cpp
    VehicleFactory(){
        //先创建两个用于克隆的原型对象
        vehicles_[Type::VEHICLE_1] = new Car("Toyota", 50.0f);
        vehicles_[Type::VEHICLE_2] = new Truck("Bens", 60.0f);
    }

    ~VehicleFactory(){
        delete vehicles_[Type::VEHICLE_1];
        delete vehicles_[Type::VEHICLE_2];
    }
    //使用克隆方法创建新对象
    Vehicle *CreateVehicle(Type type){
        return vehicles_[type]->Clone();
    }
};

//外部代码,调用 CreateVehicle 两次,克隆一个 Car 和一个 Truck
void Client(VehicleFactory &vehicle_factory){
    cout << "clone a Car\n";
    Vehicle * v1 = vehicle_factory.CreateVehicle(Type::VEHICLE_1);
    cout << "\n";
    cout << "clone a Truck \n";
    Vehicle * v2 = vehicle_factory.CreateVehicle(Type::VEHICLE_2);
    delete v1;
    delete v2;
}

int main(){
    VehicleFactory *vehicle_factory = new VehicleFactory();
    Client(*vehicle_factory);
    delete vehicle_factory;

    return 0;
}
```

10.2.5 单例方法

单例方法是一种创建设计模式,提供了一个类,该类只有一个对象,并且提供了一个全局访问点。

单例方法解决了两个重要问题:

(1)确保一个类只有一个对象,如数据库或文件。在某些场景下,我们不希望创建一个新对象,而是想得到已经创建的对象,这样做的原因是可以控制对某些共享资源的访问。

(2)提供了一个全局访问点。单例方法的好处是方便和安全,我们可以从程序中的任何位置访问对象,如访问全局变量一样,而不必担心对象会被其他代码覆盖。

单例方法的实现具有以下两个共同步骤:

(1)将默认构造函数设为私有函数,以阻止其他对象使用 new 来创建对象。

（2）定义一个静态的创建方法，如 getInstance()来充当对象的构造函数。在此静态的创建方法中调用私有构造函数来创建对象，并将其作为静态变量保存。

单例方法的 UML 类图示例如图 10.2-5 所示。

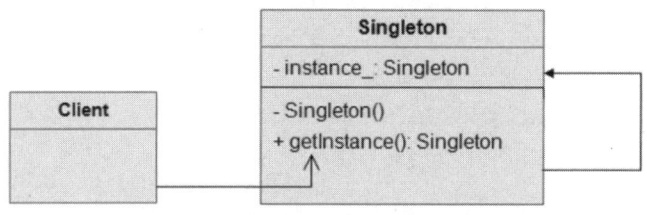

图 10.2-5　单例方法的 UML 类图示例

下面我们以 DataBase 作为对象来展示单例方法的使用，程序代码简单易读。

【例 10-5】

```cpp
#include<stream>
#include<thread>

class Database{
    //构造函数放在 protected 部分是为了阻止使用 new 来创建 DataBase 对象
protected:
    Database(const std::string connection),connection_(connection){}
    static Database* db_;
    std::string connection_;
public:
    //Database 不能被克隆
    Database(Database &db) = delete;
    //Database 也不能被赋值
    void operator=(const Database&) = delete;
    static Database *GetDatabase();          //静态函数，用于创建对象
    void SqlQuery(std::string strSQL)        //非静态函数，必须通过类的对象进行访问
    {
        cout << strSQL << endl;
    }
    static void deleteDatabase();            //静态函数用于最后释放指针
};

Database* Database::db_ = nullptr;           //静态变量必须赋初值
Database *Database::GetDatabase(const std::string connection){
    if(db_ ==nullptr){
        db_ = new Database(connection);
    }
    return db_;
}

void Database::deleteDatabase() {
    if(db_!=nullptr)
        delete db_;
        db_ = nullptr;
```

```cpp
}
//线程函数
void ThreadQuery(const std::string connection , const std::string strSQL)
{
    std::this_thread::sleep_for(std::chrono::milliseconds(1000));
    DataBase *db = DataBase::GetDatabase(connection);
    db ->SqlQuery(strSQL);
}
//全局函数
void DestroyDatabase(){
    Database::deleteDatabase();
}
int main(){
    std::string connection;
    std::string strSQL;
    connection = "connect to database";
    strSQL = "select from ...";
    std::thread t(ThreadQuery, connection, strSQL);
    t.join();
    DestroyDatabase();
    return 0;
}
```

10.3 结构设计模式

结构设计模式（structural design pattern）是软件开发中设计模式的一个子集，它侧重于类或对象的组合，以形成更大、更复杂的结构。结构设计模式有助于组织和管理对象之间的关系，从而在软件系统中实现更大的灵活性、可重用性和可维护性。结构设计模式主要包括适配器方法（adapter method）、桥接方法（bridge method）、组合方法（composite method）、装饰器（decorator method）、门面方法（facade method）、代理方法（proxy method）、蝇量级方法（flyweight method）。下面我们逐一介绍每种方法。

10.3.1 适配器方法

适配器方法是一种结构设计模式，它允许将现有类的接口用作另一个类的接口，充当两个不兼容类接口之间的桥梁，使它们协同工作。适配器方法中有一个称为适配器的类，由它负责连接独立的或不兼容的类接口。

适配器方法一般由以下四个部分组成：

（1）Client Interface：定义供用户使用的类接口，该接口由 Client 的代码调用。

（2）Adaptee：一个类或系统，它有不兼容的接口，但需要被集成到系统中。

（3）Adapter：一个类，它实现目标接口，并在内部使用 Adaptee 的实例，使它与目标接口兼容，起桥接的作用。

（4）Client：使用目标接口的代码，它不关心 Adaptee 和 Adapter 的实现细节。

适配器方法的 UML 类图示例如图 10.3-1 所示。

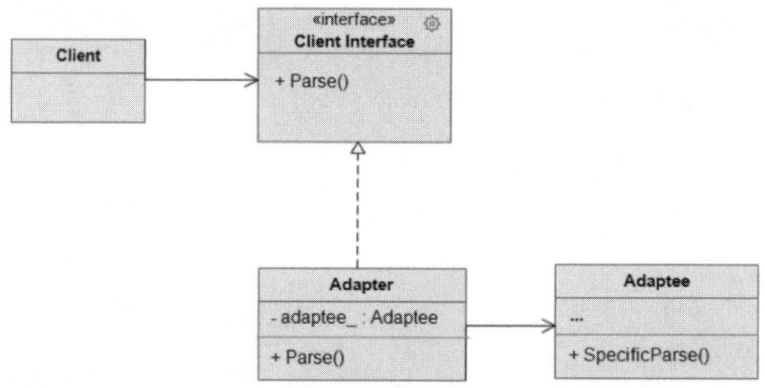

图 10.3-1　适配器方法的 UML 类图示例

下面举例说明如何使用适配器方法。假设我们正在创建一个股票分析应用程序，该应用程序首先以 XML 格式从多个数据源下载股票数据，然后为用户显示美观的图表和曲线。现在，有一个第三方的股票分析软件包，该软件包采用数据库的形式，而且仅适用于 JSON 格式的股票数据。这个第三方的股票分析软件包已经下载了大量的股票数据，但我们无法用仅支持 XML 格式的股票分析应用程序打开股票数据。为了解决格式不兼容的问题，我们创建了一个从 XML 到 JSON 的适配器。当适配器被调用时，它会将传入的 JSON 格式数据转换为 XML 格式数据。

【例 10-6】

```
#include<stream>
#include<string>

class XMLParser {
public:
    virtual ~XMLParser() = default;
    virtual std::string Parse() const override{
        return "Target interface: The default target's behavior.";
    }
};

//Adaptee，它的接口是不兼容的
class JsonParser{
public:
    std::string SpecificParse() const {
        return ".ecivreS eht fo roivaheb laicepS";
    }
};

//Adapter，它桥接 XMLParse 与 JsonParser
class XMLParserAdapter : public XMLParser{
private:
    JsonParser *jsonParser_;
public:
    XMLParserAdapter(JsonParser*jsonParser):jsonParser_(jsonParser){}
```

```cpp
        std::string Parse() const final {
            std::string info =this->jsonParser_->SpecificParse();
            std::reverse(info.begin(), info.end());
            return "Adapter: (TRANSLATED) " + info;
        }
    };

    void ClientCode(const XMLParser *xmlParser) {
        std::cout << xmlParser->Parse();
    }

    int main() {
        XMLParser *xParser = new XMLParser;
        xParser->Parse();
        std::cout << "\n\n";

        JsonParser *jPareser = new JsonParser;
        jPareser->SpecificParse();
        std::cout << "\n\n";

        XMLParserAdapter *adapter = new XMLParserAdapter(jPareser);
        ClientCode(adapter);
        std::cout << "\n";

        delete xParser;
        delete jPareser;
        delete adapter;

        return 0;
    }
```

10.3.2 桥接方法

桥接方法是一种结构设计模式，它允许用户将一个大类或一组密切相关的类拆分为两个单独的层次结构（抽象层和实现层），这两个层次结构可以被相互独立地开发。在处理跨平台应用程序、支持多种类型的数据库服务器或者某种类型的多个 API 提供商（如云平台、社交网络等）合作时，桥接方法特别有用。

在讨论桥接方法时，经常使用两个术语：实现（implementation）和抽象（abstraction）。这里的抽象和实现不是编程语言中的术语。在图 10.3-2 中，类 Remote 对应桥接方法中的抽象，类 Device 对应桥接方法中的实现。在 UML 类图中，类 Device 被定义为接口类，提供了一组接口函数（所定义的接口函数均为纯虚函数）。类 Remote 同样可以是接口类，也可以不是接口类。从类 Device 派生出来的类 Television 和类 Radio 需要实现接口类中定义的接口函数，如 isEnabled()、enable()、disable()等。抽象提供了一种高级控制逻辑，它依赖于实现对象来完成实际的工作。抽象可能会定义与实现相同的方法，但通常会声明一些复杂的行为，这些行为依赖于实现中声明的各种基元操作。在图 10.3-2 中，类 AdvancedRemote 是从类 Remote 派生而来的，将提供一些虚函数的不同实现，并桥接到具体的类 Radio 和类 Television。

桥接方法的 UML 类图示例如图 10.3-2 所示，例 10-7 给出了一个简单示例，为了节省篇幅，我们在类 Device 和类 Remote 中只定义了一个接口函数。

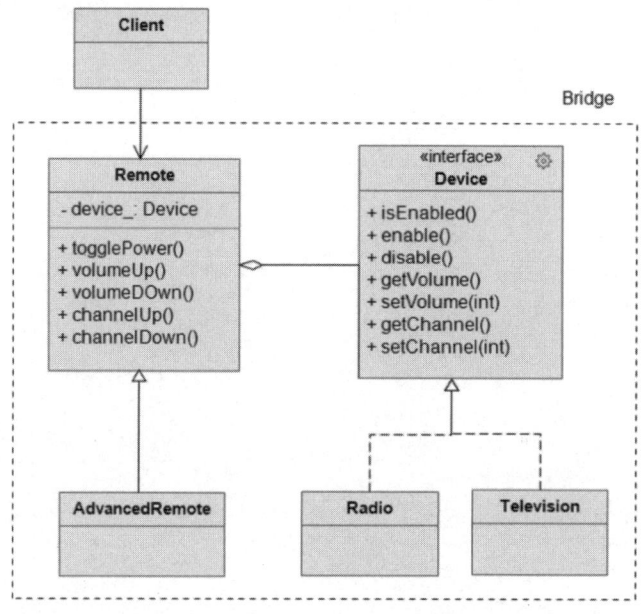

图 10.3-2　桥接方法的 UML 类图示例

【例 10-7】

```
#include<stream>
#include<string>

class Device{
public:
    virtual ~Device() {}
    virtual std::string OperationOnDevice() const = 0;
};

class Television : public Device{
public:
    std::string OperationOnDevice() const override {
        return "Television: Here's the result on TV platform. \n";
    }
};

class Radio : public Device{
public:
    std::string OperationOnDevice() const override{
        return "Radio: Here's the result on Radio platform.\n";
    }
};

class Remote{
protected:
```

```cpp
        Device* device_;
public:
        Remote(Device* device) : device_(device) {}
        virtual ~Remote() {}
        virtual std::string Operation() const override{
                return "Base operation with:\n" + this->device_->OperationOnDevice();
        }
};

class AdvancedRemote: public Remote{
public:
        AdvancedRemote(Device* device):Remote(device){}
        std::string Operation() const final {
                return "Extended operation with:\n" + this->device_->OperationOnDevice();
        }
};

void ClientCode(const Remote* remote) {
        std::cout << remote->Operation();
}

int main() {
        Device* tv = new Television();

        Remote* remote = new AdvancedRemote(tv);
        ClientCode(remote);
        std::cout << std::endl;
        delete tv;
        delete remote;

        Device* radio   = new Radio();
        remote   = new AdvancedRemote(radio);
        ClientCode(*remote);

        delete radio;
        delete remote;

        return 0;
}
```

10.3.3 组合方法

组合方法是一种结构设计模式，它允许我们将对象组合成树结构，以表达部分与整体的关系，然后像处理单个对象那样处理树结构。只有当应用的核心模型可以表示为树结构时，使用组合方法才有意义。树中的对象与组件有共同的接口函数，该接口函数供 Client 调用。

组合方法通常包括以下几个组成部分。

（1）Component：接口类，是组合方法中所有对象的通用接口，它定义了 Leaf（叶）对象和

Composite（复合）对象的通用方法。

（2）Leaf：是没有任何子项的单个对象，它实现了 Component（组件），并为各个对象提供了特定的功能。

（3）Composite：它是容器对象，用于存放 Leaf 对象以及其他 Composite 对象，它实现了 Component，并提供了用于访问容器元素的方法，如添加、删除和访问子项等。

（4）Client：负责使用 Component 接口类，不加区别地处理每个对象。

组合方法的 UML 类图示例如图 10.3-3 所示。

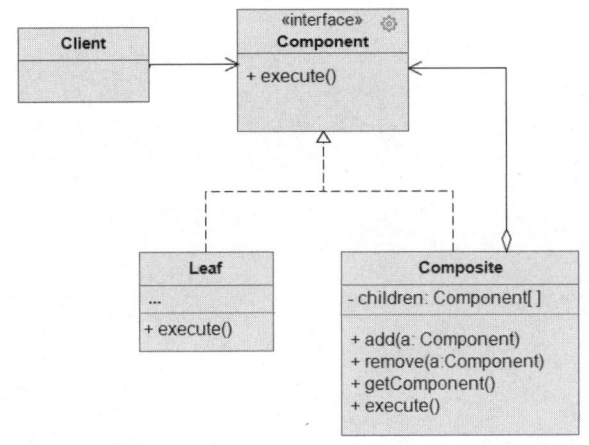

图 10.3-3　组合方法的 UML 类图示例

假设我们有两种类型的对象：产品和盒子。一个大盒子可以装入多个产品以及多个小盒子。这些小盒子也可以装入一些产品甚至更小的盒子。我们要创建和使用这样的一个订单系统：一个订单可能包含没有任何包装的简单产品，即 Leaf；也可能包含装满了产品的盒子，即 Composite。我们应该如何确定此类订单的总价格呢？

通过组合方法，我们通过一个通用接口来处理产品和盒子，该接口声明了一种计算总价的方法。如果订单是一个简单产品，则该接口只返回该产品的价格。如果订单对应于一个盒子，该接口就检查盒子里装的每件产品，查询每件产品的价格，然后返回这个盒子的总价格。如果盒子里一个子项是一个较小的盒子，即出现盒子嵌套，那么该接口也会开始检查较小盒子中的产品，依此类推，直到计算出所有内部产品的总价格。当然，盒子本身也可以有价格。组合方法的最大好处是不需要关心组成树的对象的具体类。我们不需要知道一个物体是一个简单的产品还是一个复杂的盒子，我们可以通过通用接口对它们一视同仁。调用通用接口时，对象本身会将请求传递到树中。

下面我们讨论如何设计和实现组合方法。在图 10.3-3 中，接口类 Component 向 Client 提供接口函数 execute()。类 Leaf 是一个简单类，需要实现接口函数 execute()，不过类 Leaf 的接口函数 execute()只做一些具体的事情。在 C++中，类 Leaf 是从类 Component 派生而来的。类 Composite 是一个组合类，它一方面继承了类 Component，因此同样需要提供接口函数 execute()；另一方面它也是类 Component 的聚合，即在类 Composite 中定义了一个容器，容器中的元素是 Component 对象，而且其中的某些 Component 对象可能是 Leaf 对象，另一些可能是 Composite 对象，形成了树结构。因此，在类 Composite 中定义了 add()、remove()等方法。类 Composite 中的接口函数 execute()需要遍历容器内的 Component 对象，并且只有在接口函数 execute()中才能知道容器中的元素是 Leaf 对象还是 Composite 对象。

下面我们以文件系统为例，给出一个简单的组合方法示例。

【例 10-8】

```cpp
#include <algorithm>
#include <iostream>
#include <list>
#include <string>
class FileSytemComponent{
protected:
    FileSytemComponent *parent_;
public:
    virtual ~FileSytemComponent() {}
    void SetParent(FileSytemComponent *parent){
        this->parent_ = parent;
    }
    FileSytemComponent *GetParent() const{
        return this->parent_;
    }
    virtual void Add(FileSytemComponent *component) {}
    virtual void Remove(FileSytemComponent *component) {}
    virtual bool IsDirectory() const { return false; }
    virtual std::string Execute() const = 0;
};

class File : public FileSytemComponent{
public:
    std::string Execute() const override{
        return "file ";
    }
};

class Directory : public FileSytemComponent{
protected:
    std::list<FileSytemComponent*> children_;
public:
    void Add(FileSytemComponent *component) override{
        this->children_.push_back(component);
        component->SetParent(this);
    }
    void Remove(FileSytemComponent *component) override{
        children_.remove(component);
        component->SetParent(nullptr);
    }
    bool IsDirectory() const override{
        return true;
    }
    std::string Execute() const override{
        std::string result;
```

```cpp
            for(const FileSytemComponent *c : children_){
                result += c->Execute();
            }
            return "(" + result + ")";
        }
};

void ClientCode(FileSytemComponent * f){
    //...
    std::cout << "RESULT: " << f->Execute();
    //...
}

void ClientCode2(FileSytemComponent *f1, FileSytemComponent *f2){
    if (f1->IsComposite()) {
        f1->Add(f2);
    }
    std::cout << "RESULT: " << f1->Execute();
}

int main() {
    FileSytemComponent *f0 = new File;
    std::cout << "Client: I've got a simple component:\n";
    ClientCode(f0);
    std::cout << "\n\n";

    FileSytemComponent *d0 = new Directory;
    FileSytemComponent *d1 = new Directory;

    FileSytemComponent *f1 = new File;
    FileSytemComponent *f2 = new File;
    FileSytemComponent *f3 = new File;
    d1->Add(f1);
    d1->Add(f2);
    FileSytemComponent *d2 = new Directory;
    d2->Add(f3);
    d0->Add(d1);
    std::cout << "Client: Now I've got a composite tree:\n";
    ClientCode(d0);
    std::cout << "\n\n";

    ClientCode2(d0, d2);
    std::cout << "\n";

    delete f0;
    delete d0;
    delete d1;
    delete d2;
    delete f1;
```

```
        delete f2;
        delete f3;

        return 0;
}
```

10.3.4 装饰器方法

装饰器方法是一种结构设计模式，它可以动态地为某个对象增加新行为，而不影响由同一个类生成的其他对象。装饰器方法通过创建一个装饰器类集，可以把所有的具体对象放入其中。

装饰器方法主要用于扩展功能、多种功能组合、遗留代码集成、GUI 组件和输入/输出流等场合。

装饰器方法通常包括以下几部分组成：

（1）Component：这是一个接口类，用于定义具体组件和装饰器的通用接口，它指定了可以对对象进行的操作。

（2）ConcreteComponent：是实现组件接口的具体对象或类，它们是要添加新行为或新功能的对象。

（3）Decorator：这是一个接口类，它也实现了 Component 接口，并具有对 Component 对象的引用。Decorator 负责向包装中的 Component 对象添加新行为。ConcreteDecorator 是扩展的装饰器类，它们向组件添加特定行为或功能。每个 ConcreteDecorator（如类 SMS、类 Facebook、类 Slack）都可以向组件添加一个或多个行为。

（4）Client：客户端，接收 Component 对象，调用其接口。

装饰器方法的 UML 类图示例如图 10.3-4 所示。

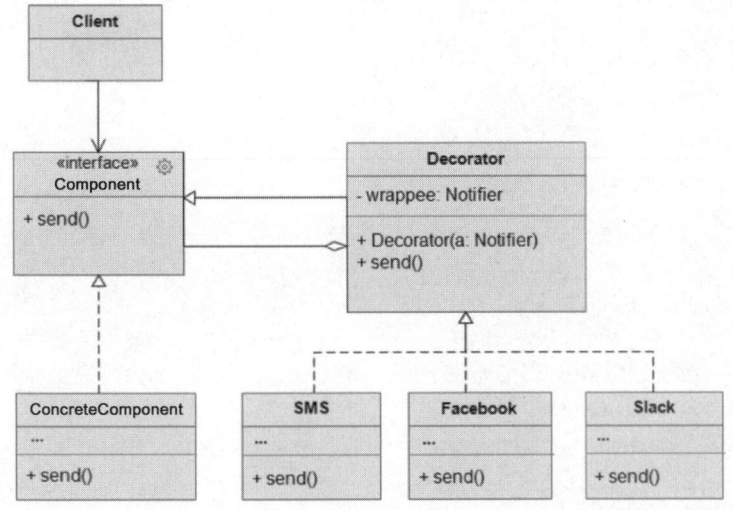

图 10.3-4　装饰器方法的 UML 类图示例

我们使用 Refactoring.Guru 网站的一个例子。假设我们正在开发一个通知库，该库允许其他程序向其用户通知重要事件的信息。该库的初始版本基于类 Notifier，该类只有几个字段、一个构造函数和一个发送方法。类 Notifier 可以接收来自 Client 的消息参数，并通过其构造函数，以电子邮件的方式将消息发送给被通知的应用程序。

在某些时候，我们意识到用户期望的不仅仅是电子邮件，许多客户希望收到重要事件的 SMS 短信，还有些客户希望在 Facebook 上收到通知，而企业用户则希望收到 Slack 通知。因此我们必须扩展类 Notifier，将类 Notifier 作为父类，从它派生出每一种通知方法的通知子类，然后由 Client 实例化相应的通知子类，将它用于发送相应的通知。

我们又收到新的需求：有用户希望收到 SMS+Facebook 的组合通知，有的用户则希望收到 SMS+Slack 的组合通知，而有的用户可能希望收到 SMS+Facebook+Slack 的组合通知，等等。

要解决上面的问题，除了继承（inheritance），我们还需要聚合（aggregation）。首先，类 Notifier 就是装饰器方法中的类 Component，定义了接口函数 send()；其次，我们设计的 Decorator 不仅是装饰器，也是一个包装器，将类 Notifier 放入其中。包装器要实现被包装对象的接口函数 send()。这样从 Client 来看，这些对象是相同的，包装器接收遵循该接口函数的任何对象，这允许我们将多个不同包装器（图 10.3-4 中的三个 ConcreteDecorator，即类 SMS、类 Facebook 和类 Slack）增加到系统中。类 SMS、类 Facebook 和类 Slack 都是 Decorator 的派生类，而包装器 Decorator 与类 Notifier 之间既是继承关系又是聚合关系。

【例 10-9】

```cpp
#include <iostream>
#include <string>

class NotifierComponent{
public:
    virtual ~NotifierComponent() {}
    virtual std::string send() const = 0;              //纯虚函数
};

class ConcreteComponent : public NotifierComponent{
public:
    std::string send() const final {
        return "Concrete Component";
    }
};

class Decorator : public NotifierComponent{
protected:
    NotifierComponent* component_;
public:
    Decorator(NotifierComponent* nc):component_(nc) {}
    std::string send() const override {
        return this->component_->send();
    }
};

class SMSDecorator : public Decorator{
public:
    SMSDecorator(NotifierComponent* nc):Decorator(nc) {}
    std::string send() const final{
        return "SMSDecorator(" + Decorator::send() + ")";
    }
```

```cpp
};

class FacebookDecorator : public Decorator{
public:
    FacebookDecorator(NotifierComponent* nc) : Decorator(nc) {}
    std::string send() const final {
        return "FacebookDecorator(" + Decorator::send() + ")";
    }
};

class SlackDecorator : public Decorator{
public:
    SlackDecorator(NotifierComponent* nc) : Decorator(nc) {}
    std::string send() const final {
        return "SlackDecorator(" + Decorator::send() + ")";
    }
};

void ClientCode(NotifierComponent* nc){
    std::cout << "RESULT: " << nc->send();
}

int main(){
    NotifierComponent* simple = new ConcreteComponent;
    std::cout << "Client: I've got a simple component:\n";
    ClientCode(simple);
    std::cout << "\n\n";
    NotifierComponent* d1 = new SMSDecorator(simple);
    NotifierComponent* d2 = new FacebookDecorator(simple);
    std::cout << "Client: Now I've got decorated components:\n";
    ClientCode(d1);
    ClientCode(d2);
    std::cout << "\n";

    delete simple;
    delete d1;
    delete d2;
    return 0;
}
```

10.3.5 门面方法

门面方法是一种结构设计模式，它为库、框架或其他复杂子系统提供了一个简化的接口函数。门面方法隐藏了底层系统的复杂性，Client 可以使用类 Facade 提供的简单接口函数与系统进行交互。

门面方法通常包括以下三个组成部分。

（1）Façade：为 Client 提供可使用的一个简化接口函数，该接口函数隐藏了子系统的复杂性。

（2）Sub_Systems：一组具有不同功能的子系统，子系统对类 Facade 毫不知情。

（3）Client：通过了 Facade 使用子系统提供的功能。

门面方法的 UML 类图示例如图 10.3-5 所示。

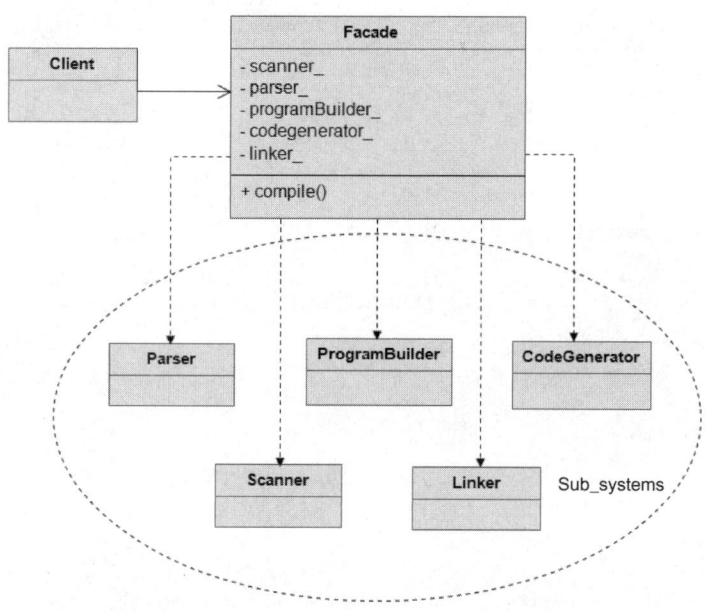

图 10.3-5　门面方法的 UML 类图示例

Facade 是一个类，它为许多复杂子系统提供简化的接口函数，每个子系统可能都有自己的接口函数，但类 Facade 提供的是更高层面的接口函数。例如，当我们需要将应用程序与具有数十种功能的复杂库集成时，但又只需要复杂库的一小部分功能时，使用门面方法将非常方便。例如，将一个搞笑短视频上传到社交网站时，上传应用程序可能会使用专业的视频转换库。但实际上，上传应用程序真正需要的只是一个类 Facade，该类有一个编码函数（文件名、格式），该编码函数与视频转换库连接。再如，当我们打电话到一家电商店下订单时，接线员就是我们通往商店所有部门和服务的门面，接线员将为我们提供订购系统、支付网关和各种交付服务的简单语音界面。

图 10.3-5 所示的 UML 类图是以编译器（Compiler）基础的门面方法 UML 类图，编译器包括词法分析、语法分析、中间代码生成、中间代码优化、目标代码生成和链接等，Facade 为 Client（如程序员）提供一个简化的接口函数，代码如例 10-10 所示。

【例 10-10】

```
#include <iostream>
#include <string>

class Parser {
public:
    Parse(){
        //进行解析;
    }
private:
    void parseExpr(){
```

```cpp
        //表达式解析
    }
};

class CodeGeneartor{
public:
    void CodeGenrate(){
        //生成代码
    }
private:
    void Optimization(){
        //代码优化
    }
};

class CompilerFacade {
    protected:
    Parser *parser_;
    CodeGeneartor *codegenerator_;
public:
    Facade(Parser *parser = nullptr, CodeGeneartor *codegenerator = nullptr) {
        parser_ = parser? parser: new parser;
        codegenerator_ = codegenerator ? codegenerator: new CodeGeneartor;
    }
    ~CompilerFacade(){
        delete parser_;
        delete codegenerator_;
    }
    //CompilerFacade 提供对外的接口函数 Compile()
    void Compile(){
        this->parser->Parse();
        this->codegenerator_->CodeGenerate();
    }
};

//Client 调用类 CompilerFacade 提供的简单接口函数 Compile()
void ClientCode(CompilerFacade *facade){
    facade->Compile();
}

int main(){
    Parser *p = new Parser;
    CodeGeneartor *cg = new CodeGeneartor;
    Facade *facade = new CompilerFacade(p, cg);
    ClientCode(facade);
    //这里不用销毁 p 和 cg，在销毁 Facade 对象时将销毁它们
    delete facade;
    return 0;
}
```

10.3.6 代理方法

代理方法是一种结构设计模式，用于为一个对象提供替代或占位，目的是控制对该对象的访问，因为代理能在访问到达前或者访问后进行有关操作，以改变访问结果。

微信和支付宝都是银行账户的代理，两者都实现了相同的接口：它们可用于付款和收款。消费者感觉很好，因为没有必要去银行取现金随身携带。店主也很高兴，因为交易收入是以电子方式添加到商店的银行账户中的，节省了去银行存款的时间，更没有了在去银行路上被偷或者被抢劫的风险。微信和支付宝都收取费用或者提供优惠，这就是一种访问到达前或者访问后进行有关操作的例证。

通过代理方法，我们可以创建一个与原始服务对象具有相同接口的新代理类，然后更新应用，以便将代理对象传递给原始对象的所有 Client，一旦收到 Client 的请求，代理对象就会创建一个真实的服务对象，并将所有工作都委托给服务对象。这样原本在 Client 创建原始服务对象的初始化代码便移到代理对象的内部了。

代理方法通常包括以下几个部分组成：

（1）Subject：是一个接口类，用于定义 RealSubject 对象和 Proxy 对象的共享公共接口，声明了 Proxy 对象用于控制对 RealSubject 对象访问的方法。

（2）RealSubject：是 Proxy 对象所代表的实际对象，它包含业务逻辑或 Client 代码要访问的资源，实现了 Subject 声明的操作，表示由 Proxy 对象控制访问的实际资源或对象。

（3）Proxy：RealSubject 对象的代理，它控制对真实对象的访问，并可能提供其他功能，如延迟加载、访问控制或日志记录。

（4）Client：访问 Subject 的 Client 代码。

图 10.3-6 所示的示例是一个基于微信移动支付的代理方法的 UML 类图。类 MobilePay 是 Subject，是一个接口类，为 Client 提供了三个接口函数，即 receive()、redPacket()、transfer()。在 C++中，这三个接口函数都是纯虚函数。类 WeChatPayProxy 是代理类，是从类 MobilePay 派生出来的，同时它将 SavingAccount 作为自己的成员，其接口函数 receive()、redPacket()、transfer()用于代理 SavingAccount 的收支。SavingAccount 是提供真正的服务，它要实现三个接口函数，即 receive()、redPacket()、transfer()，这三个接口函数是通过它的两个私有函数 save()和 withdraw()来实现的。

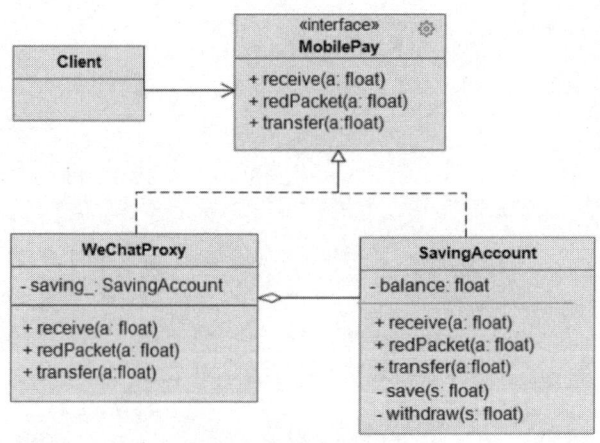

图 10.3-6 代理方法的 UML 类图示例

【例 10-11】

```cpp
#include <iostream>
#include <string>

class MobilePay{
public:
    virtual void receive(float a) = 0;
    virtual void redPacket(float a) = 0;
    virtual void transfer(float a) = 0;
};

class SavingAccount : public MobilePay{
public:
    void receive(float a) { save(a);}
    void redPacket(float a) { withdraw(a);}
    void transfer(float a) { withdraw(a); }
private:
    void save(float a) { balance += a; }
    void withdraw(float a) { balance -= a; }
    float balance = 0.0;
};
class WeChatPayProxy : public MobilePay{
private:
    SavingAccount *saving_;
public:
    //代理一个原始服务对象
    Proxy(SavingAccount *saving):saving_(new SavingAccount (*saving)){}
    ~Proxy(){
        delete saving_;
    }
    void receive(float a){
        this->saving_->receive(a);
    }
    void redPacket(float a) {
        this->saving_->redPacket(a);
    }
    void transfer(float a) {
        this->saving_->transfer(a);
    }
};

void ClientPay( MobilePay* pay, float a){
    pay->transfer(a);
}

void ClientReceive(MobilePay* recv, float a){
    recv->receive(a);
}
```

```cpp
void ClientRedPacket(MobilePay* pay, float a){
    pay->redPacket(a);
}

int main(){
    SavingAccount *saving = new SavingAccount;
    ClientReceive(saving, 100);

    WeChatPayProxy *proxy = new WeChatPayProxy (saving);
    ClientPay(proxy, 60);
    ClientRedPacket(proxy, 40);

    delete saving;
    delete proxy;
    return 0;
}
```

10.3.7 蝇量级方法

蝇量级方法是一种结构设计模式，通过共享多个对象状态之间的公共部分，而不是将所有的数据都保留在每个对象中，可以使我们将更多的对象放入有限的内存中。在使用蝇量级方法前，请确保程序中存在大量类似对象消耗内存的问题。

当我们需要创建数以万计的非常相似的对象时，选择蝇量级方法是一个明智的决定。蝇量级方法中的对象的一个重要特征是它们是不可变的，这意味着它们一旦创建后就不能被修改。

假如我们需要编写一个烟花程序。程序需要在庆祝元旦时模拟放烟花，当烟花升空后爆炸，将产生数以万计的绚丽的烟花粒子。每个烟花粒子类 Particle 都有颜色字段和精灵字段。这两个字段在所有烟花粒子中存储的数据几乎相同。烟花粒子状态的其他部分，如坐标、运动向量和速度，对于每个烟花粒子来说都是唯一的。这些字段的值会随着时间而变化，而每个烟花粒子的颜色和精灵在爆炸期间保持不变。

蝇量级方法有两个重要的概念：

（1）内在状态（intrinsic state）：我们通常称对象的常量数据为内在状态，它存在于物体中，其他对象只能读取它，而不能更改它。

（2）外在状态（extrinsic state）：对象状态的其余部分，经常被其他物体"从外部"改变。

在使用蝇量级方法时，建议不要在对象的内部存储外在状态。相反，应该将外在状态传递给依赖于它的特定方法。只有内在状态保留在对象中，允许在不同的上下文中重用它，这样会减少所需要的对象总数，因为它们仅在内在状态上有所不同，而内在状态的变化比外在状态少得多。

蝇量级方法通常包括以下几个部分组成：

（1）Flyweight：类 Flyweight 包含可在多个对象之间共享的原始状态部分，存储在类 Flyweight 中的状态称为内在状态，传递给类 Flyweight 的方法的状态称为外在状态。

（2）FlyweightFactory：管理现有的 Flyweight 对象。在 FlyweightFactory 中，Client 不会直接创建类 Flyweight 的对象。客户会调用 FlyweightFactory，向它传递类 Flyweight 所需的内在状态。FlyweightFactory 会查看以前创建的 Flyweight 对象，并返回与搜索条件匹配的现有的 Flyweight 对象。如果未找到任何内容，则创建一个新的 Flyweight 对象。

（3）Context：包含外在状态，外在状态在所有原始对象中都是唯一的。当类 Context 与某一个 Flyweight 对象配对时，它表示原始对象的完整状态。

（4）Client：计算或存储类 Flyweight 的外在状态。从 Client 的角度来看，Flyweight 对象是一个模板对象，因此可以在运行时对其进行配置。

蝇量级方法的 UML 类图示例如图 10.3-7 所示。

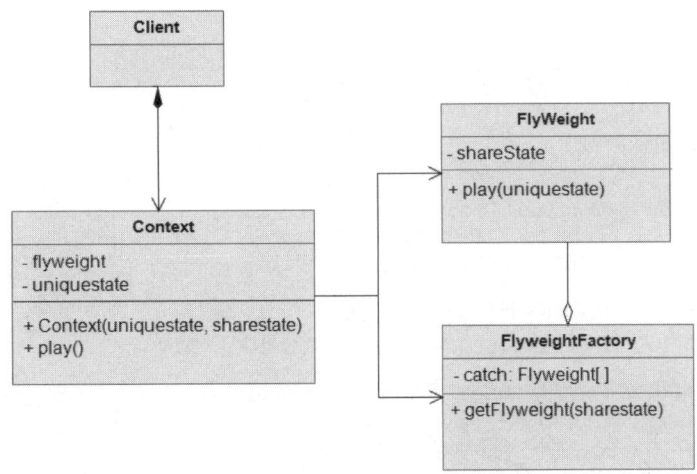

图 10.3-7　蝇量级方法的 UML 类图示例

采用蝇量级方法设计烟花程序的示例代码如下。

【例 10-12】

```cpp
#include <algorithm>
#include <iostream>
#include <vector>
#include <string>

struct ParticleSharedState{
    ParticleSharedState(const std::string &color, const int size) : color_(color), size_(size) {}
    std::string color_;                    //烟花粒子的颜色
    int radius_;                           //烟花粒子的半径
};

struct ParticleUniqueState{
    float positionX_, positionY_, positionZ_;      //烟花粒子的空间位置
    float speedX_, speedY_, speedZ_;               //烟花粒子在 X、Y、Z 方向上的速度

    ParticleUniqueState(const float positionX, const float positionY, const float positionZ,
                const float speedX, const float speedY, const float speedZ): positionX_(positionX),
                positionY_(positionY), positionZ_(positionZ), speedX_(speedX),
                speedY_(speedY), speedZ_(speedZ) {}
};

//ParticleFlyweight 存储内在状态，也接收外在状态
class ParticleFlyweight{
```

```cpp
private:
    ParticleSharedState *shared_state_;              //只保留内在状态
public:
    ParticleFlyweight(const ParticleSharedState *shared_state)
                :shared_state_(new ParticleSharedState(*shared_state)) {}

    ParticleFlyweight(const ParticleFlyweight &other)
                :shared_state_(new ParticleSharedState(*other.shared_state_)){}

    ~ParticleFlyweight(){
        delete shared_state_;
    }
    ParticleSharedState *getParticleSharedState() const{
        return shared_state_;
    }
    void play (const ParticleUniqueState &unique_state){
        //这里的代码是用 shared_state_ 和 unique_state 配对成一个烟花粒子
    }
};

class ParticleFlyweightFactory{
private:
    std::vector<ParticleFlyweight> particle_flyweights_;
public:
    ParticleFlyweightFactory(std::vector<ParticleSharedState> share_states){
        for (const ParticleSharedState &pss : share_states){
            this->particle_flyweights_.push_back(pss);
        }
    }
    ParticleFlyweight GetParticleFlyweight(int idx){
        return this->particle_flyweights_[idx];
    }
    int getParticleFlyweightsCount(){
        return this->particle_flyweights_.size();
    }
};

void AddParticleFireworks( ParticleFlyweightFactory &pff, const ParticleUniqueState &pus ){
    std::cout << "\nClient: Adding a particle to fireworks.\n";
    size_t count = pff.getParticleFlyweightsCount();
    int idx = random(count);
    const ParticleFlyweight & particleflyweight = pff.GetParticleFlyweight( idx );
    particleflyweight.play( pus);
}

int main(){
    std::vector <ParticleSharedState> &pss = {{ "red", 0.001}, { "green", 0.001}, { "blue", 0.001}}
    ParticleFlyweightFactory *factory = new ParticleFlyweightFactory(pss);
```

```
        ParticleUniqueState &pus0 {20.0, 10.0, 20.0, 15.0, 10.0, 15.0};
        AddParticleFireworks(*factory, pus0);

        ParticleUniqueState &pus1 {10.0, 20.0, 10.0, 15.0, 20.0, 20.0};
        AddParticleFireworks(*factory, pus1);

        ParticleUniqueState &pus2 {20.0, 20.0, 10.0, 20.0, 10.0, 20.0};
        AddParticleFireworks(*factory, pus2);

        delete factory;
        return 0;
    }
```

10.4 行为设计模式

行为设计模式（behavioral design pattern）是软件开发中设计模式的一个子集，用于处理对象和类之间的交互，侧重于对象和类如何协作与通信，从而完成任务和职责。行为设计模式主要包括责任链方法（chain of responsibility method）、迭代器方法（iterator method）、中介器方法（mediator method）、备忘录方法（memento method）、观察者方法（observer method）、状态方法（state method）、策略方法（strategy method）、模板方法（template method）、命令方法（command method）、访客方法（visitor method）。

10.4.1 责任链方法

责任链方法是一种行为设计模式，用于在程序链中的传递请求。收到请求后，每个处理程序决定是处理响应请求，还是将响应请求传递给程序链中的下一个处理程序。

责任链方法依赖于将特定行为转换为独立对象的程序，并称之为处理程序。采用责任链方法时，建议将这些处理程序链接到一个程序链中。每个处理程序都有一个字段，用于存储对程序链中下一个处理程序的引用。除了处理请求，处理程序还会沿程序链进一步传递响应请求，以便所有的处理程序都有机会处理响应程序。

责任链方法的 UML 类图示例如图 10.4-1 所示，带有<<interface>>的类 Button 是处理程序，它是一个接口类，声明了的两个接口函数，即 Click()和 passToNext()。这两个接口函数都是纯虚函数，需要在派生类中实现。passToNext()用于在程序链中将请求传递给下一个处理程序。BaseButtonHandler 是基类，也是派生类，它要实现接口函数 Click()和 passToNext()，同时把类 Button 作为自己的成员。类 CheckboxHandler 和类 RadioButtonHandler 是从类 BaseButtonHandler 派生出来的，是真正响应请求的处理程序，即 ConcreteHandler，因此要实现处理接口函数 Click()和 passToNext()。我们注意到：类 BaseButtonHandler 和类 Button 的关系既是实现关系，又是聚合关系，而类 BaseButtonHandler 和类 CheckboxHandlers、类 RadioButtonHandler 之间是泛化关系。在下面的 C++示例代码中，BaseButtonHandler 对应着矩形按钮。

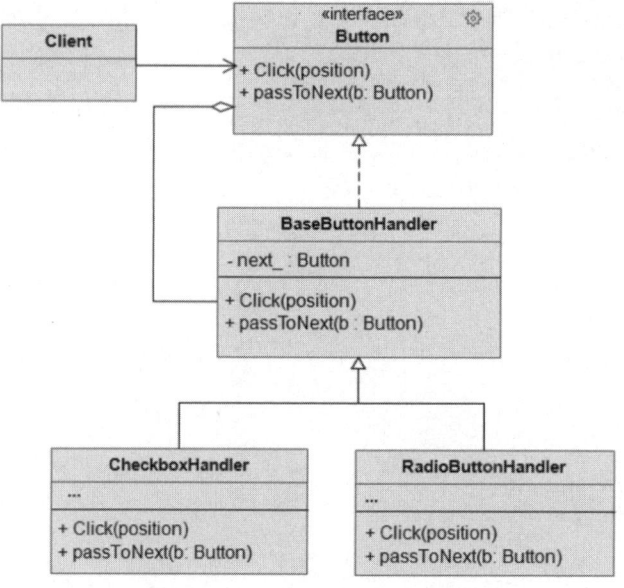

图 10.4-1　责任链方法的 UML 类图示例

【例 10-13】

```
#include <iostream>
#include <string>

class Button {
public:
    virtual Button *passToNext(Button *button) = 0;
    virtual void Click(int position) = 0;
};

//BaseButtonHandler 对应着矩形按钮
class BaseButtonHandler : public Button {
protected:
    Button *next_;
public:
    BaseButtonHandler():next_(nullptr) {}
    Button *passToNext(Button *next) override {
        this->next_ = next;
        return next_;
    }

    virtual void Click(int position) override {
        if (ClickWithinButton(position)){
            std::cout << "Normal button is clicked."
        }
        else{
            if(this->next_)
                this->next_->Click(position);
        }
```

```cpp
    }

    virtual bool ClickWithinButton(int position)override{
        if (position == 0)
            return true;
        else
            return false;
    }
};

//ConcreteHandler 是具体的处理程序
class CheckboxHandler : public BaseButtonHandler {
public:
    void Click(int position) final {
        if (ClickWithinButton(position)) {
            std::cout << " the checkbox is checked.";
        }
    }

    bool ClickWithinButton(int position)final{
        if (position == 1)
        return true;
        else
        return false;
    }
};

//ConcreteHandler 是具体的处理程序
class RadioButtonHandler : public BaseButtonHandler {
public:
    void Click(int position) final {
        if (ClickWithinButton(position)) {
            std::cout << "the radiobutton is clicked.";
        }
    }

    bool ClickWithinButton(int position)final{
        if (position == 2)
        return true;
        else
        return false;
    }
};

void ClientClickButton(BaseButtonHandler* button, int position) {
    button->Click(position);
}

int main() {
```

```cpp
        BaseButtonHandler * button = new BaseButtonHandler;
        CheckboxHandler   *checkbox = new CheckboxHandler;
        ClientClickButton(button, 0);
        ClientClickButton(checkbox, 1);

        Button->passToNext(checkbox);
        ClientClickButton(button, 1);

        RadioButtonHandler *radioButton = new RadioButtonHandler;
        ClientClickButton(radioButton, 2);

        checkbox->passToNext(radioButton);
        ClientClickButton(checkbox, 2);

        delete button;
        delete checkbox;
        delete radioButton;
        return 0;
}
```

10.4.2 迭代器方法

迭代器方法是一种行为设计模式，用于遍历集合的元素，且无须公开其基础表示形式（如列表、堆栈、树等）。

集合是一组对象的容器，是编程中最常用的数据类型之一。大多数集合是基于简单列表数据结构的，也有一些是基于堆栈、树、图形和其他复杂数据结构的。但是，无论集合的数据结构如何，它都必须提供某种访问其元素的方法，以便其他代码可以使用这些元素。访问集合的方法应能够遍历集合的每个元素，且不会访问相同的元素。

迭代器方法通常包括以下几个组成部分：

（1）Iterator：定义了用于访问和遍历集合元素的接口函数，如 first()、hasNext()、next()等。

（2）ConcreteIterator：用于实现具体的迭代器，实现了类 Iterator 的接口函数，并在遍历集合时保持当前位置。

（3）IterableCollection：定义了用于创建 Iterator 对象的接口函数，通常包括一个类似于CreateIterator()的方法，该方法返回集合的 Iterator 对象。

（4）ConcreteCollection：用于实现具体的接口函数，表示对象的集合，提供了用于创建可遍历集合元素的 Iterator 对象的具体实现。

（5）Client：通过 IterableCollection 创建 ConcreteIterator，并通过 Iterator 的接口函数遍历集合中的元素。

迭代器方法的主要思想是将集合的遍历行为提取到一个称为迭代器的单独对象中。除了实现算法本身，迭代器对象还封装了所有的遍历细节，例如，在集合的当前位置，以及到最后还剩下多少元素。通常，Client 使用迭代器访问集合元素，直到它不返回任何内容为止，这意味着迭代器已遍历了集合的所有元素。所有迭代器必须实现相同的接口函数，使得 Client 代码可以与任何集合类型或任何遍历算法兼容。

图 10.4-2 所示为迭代器方法 UML 类图示例。图中 Iterator 是接口类，声明了遍历集合元素所

需的操作,如获取第一个元素的接口函数是 first()、获取下一个元素的接口函数是 next()、检索当前位置的接口函数是 current()等。在 C++中,Iterator 是抽象类,其接口函数都被定义为纯虚函数。ConcreteIterator 是具体的迭代器,是从类 Iterator 派生出来的,要实现类 Iterator 中的所有接口函数,这些接口函数用于遍历集合元素,并自行跟踪遍历进度。这种特性允许多个迭代器彼此独立地遍历同一个集合的元素。IterableCollection 是抽象类,声明了一种或多种接口函数,用于获取与集合兼容的迭代器。请注意,接口函数 CreateIterator()的返回类型必须声明为迭代器接口类型,即 Iterator,以便集合可以返回对应的迭代器。在 Client 请求创建一个特定的迭代器类时,类 ConcreteCollection 会返回一个具体迭代器的实例。类 ConcreteCollection 是从类 IterableCollection 派生出来的。Client 通过集合和迭代器的接口函数处理业务,这样 Client 就不会耦合到具体的类,从而可以使用相同的 Client 对接各种集合和迭代器。通常,Client 不会自行创建迭代器,而是从集合中获取迭代器,只是在某些情况下,Client 才直接创建一个特殊迭代器。

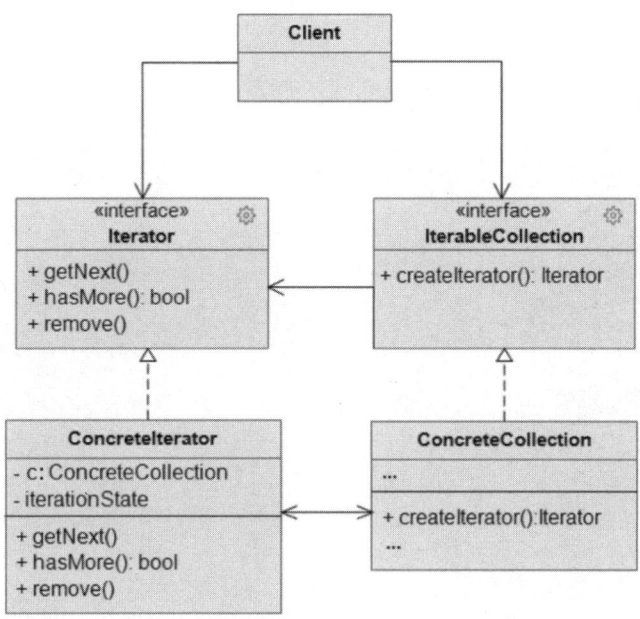

图 10.4-2　迭代器方法的 UML 类图示例

【例 10-14】

```
#include <iostream>
#include <string>
#include <vector>
using namespace std;

//模板接口类 Iterator
template <typename T>
class Iterator {
public:
    T* First()       =0;
    T* void Next()   =0;
    bool hasNext()   =0;
    T* Current()     =0;
};
```

```cpp
//数据对象
class Employee {
private:
    string name_;
    double salary_;
public:
    Employee(string name, double salary) {
        name_ = name;
        salary_ = salary;
    }
    double getSalary() { return salary_; }
    string getName() { return name_; }
};

//具体的迭代器
template <class T>
class ConcreteIterator: public  Iterator<T>{
public:
    using iter_type = std::vector<T>::iterator ;
    ConcreteIterator(Container<T>&p_data, bool reverse = false) : m_p_data_(p_data){
        m_p_data_ = p_data;
        m_it_ = m_p_data_.m_data_.begin();
    }
    T* First(){
        m_it_ = m_p_data_.m_data_.begin();
        return m_it_;
    }
    T* Next() { return m_it_++; }
    bool hasNext(){ return (m_it_ < m_p_data_.m_data_.end());}
    T* Current() { return m_it_;}

private:
    Container<T>& m_p_data_;
    iter_type m_it_;
};

//模板接口类 Container
template <class T>
class Container{
public:
    //创建 Iterator 对象的接口函数
    virtual Iterator<T> * CreateIterator() = 0 ;
};

//具体的容器类
template <class T>
class ConcreteContainer : public Container <T> {
public:
```

```cpp
        Iterator<T> * CreateIterator() {
            return new ConcreteIterator<T> (*this);
        }
        void Add(T a){ m_data_.push_back(a);}
        std::vector<T> m_data_;
};

void ClientCode(Container* company){
    Iterator<Employee>* it = company->CreateIterator();
    for (it->First(); it->hasNext(); it->Next()) {
        std::cout << it->Current()->getName() << ","
            << it->Current()->getSalary() << std::endl;
    }
}

int main(){
    ConcreteContainer<Employee>* company = new ConcreteContainer<Employee>();
    Company->Add("Join", 80000.0);
    Company->Add("Tom", 90000.0);
    Company->Add("Smith", 100000.0);

    ClientCode(company);
    delete company;
    return 0;
}
```

10.4.3　中介器方法

中介器方法是一种行为设计模式，定义了一个对象（中介器），以集中系统中各种组件或对象之间的通信。通过阻止组件之间的直接通信，并让它们通过中介器进行通信来降低耦合，可以为系统架构带来更好的可维护性和灵活性。

中介器方法是重要且广泛使用的行为设计模式之一。中介器通过在两个对象之间引入一个层来实现对象之间的解耦，以便对象之间的通信通过该层完成。在现实生活中，一个大型机场每天都有成百上千架飞机起降，而起降的跑道只有三五条。为了避免飞机起降时发生碰撞，必须由塔台来指挥飞机的起降，而不是让起降飞机之间直接通信。这样的空中交通指挥控制中心实际上就是一个中介器。另一个例子是我们在编程时的对话框（Dialog），Dialog 中有许多个控件，如按钮（Button）、编辑框（Edit）、复选框（Checkbox）、单选框（RadioButton）、组合框（CommboBox）等，这些控件彼此并不直接互相通信，而是通过 Dialog 进行通信的。

中介器方法通常包括以下几个组成部分：

（1）Mediator：定义了接口函数，指定了 ConcreteMediator 应实现的接口函数，封装了用于协调和管理对象之间通信的逻辑，实现了对象之间的松耦合，并集中对对象进行控制。

（2）Colleague：定义了相互通信的组件或对象，这些组件或对象通过类 Mediator 的接口函数进行通信，每个类 Colleague 只知道类 Mediator，而不知道其他的类 Colleague。这种隔离可以确保对某个类 Colleague 的更改不会直接影响其他的类 Colleague。

（3）ConcreteMediator：实现了类 Mediator 的具体接口函数，用于协调 ConcreteColleague 之间的通信，处理 ConcreteColleague 之间的通信并确保组织良好的协作，同时保持 ConcreteColleague 之间的解耦。

（4）ConcreteColleague：实现了具体的 Colleague 接口函数，ConcreteColleague 通过类 Mediator 与其他的类 Colleague 进行通信，避免了类 Colleague 之间的直接依赖，使系统架构更灵活和更具可维护性。

（5）Client：是类 Mediator 接口函数的调用者。

图 10.4-3 所示为中介器方法的 UML 类图示例。图中，类 Airplane 对应于类 Colleague；类 CommercialAirplane 对应于类 ConcreteColleague，实现了三个接口函数；类 ControlTower 是接口类，对应于类 Mediator，定义了两个接口函数，即 requestTakeoff()和 requestLanding()，这两个接口函数在类 AirportControlTower 中实现。在 C++中，类 AirportControlTower 是从类 ControlTower 派生出来的。类 AirportControlTower 中的 requestTakeoff()和 requestLanding()用于处理请求或者应答请求。

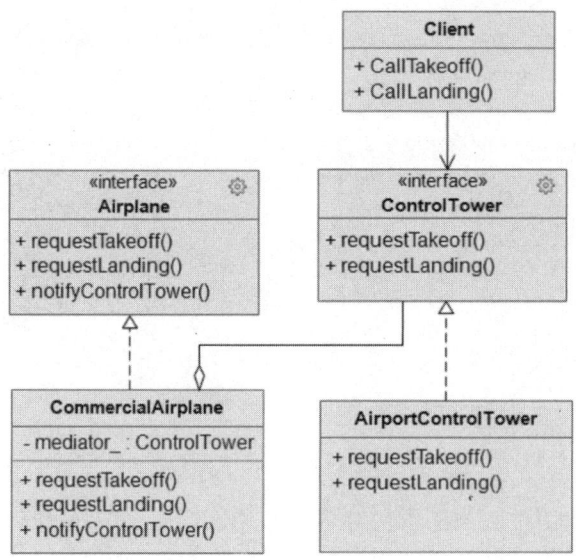

图 10.4-3　中介器方法的 UML 类图示例

【例 10-15】

```
#include<iostream>
#incldue<string>

class ControlTower;
//Colleague
class Airplane{
public:
    virtual void requestTakeoff() = 0;
    virtual void requestLanding() = 0;
    virtual void notifyAirTrafficControl(std::string message) = 0;
};

//ConcreteColleague
```

```cpp
class CommercialAirplane : public Airplane {
public:
CommercialAirplane(ControlTower* mediator) {
                mediator_ = mediator;
}

void requestTakeoff() {
    mediator_->requestTakeoff(this);
}

void requestLanding() {
    mediator_->requestLanding(this);
}

void notifyAirTrafficControl(std::string message) {
    cout<<"Commercial Airplane: " + message);
}
private:
    ControlTower* mediator_;
};

//ConcreteColleague
class MilitaryAirplane : public Airplane {
public:
    MilitaryAirplane(ControlTower* mediator) {
        mediator_ = mediator;
    }
    void requestTakeoff() { mediator_->requestTakeoff(this);}
    void requestLanding() { mediator_->requestLanding(this);}
    void notifyAirTrafficControl(std::string message) {
        cout<<"Military Airplane: " + message);
    }

private:
    ControlTower* mediator_;
};

//Mediator 的接口函数
class ControlTower {
    void requestTakeoff(Airplane airplane);
    void requestLanding(Airplane airplane);
};

//ConcreteMediator
class AirportControlTower : public ControlTower{
public:
    void requestTakeoff(Airplane* airplane){airplane->
        notifyAirTrafficControl("Requesting takeoff clearance.");}
```

```cpp
            void requestLanding(Airplane* airplane) {airplane->
                        notifyAirTrafficControl("Requesting landing clearance.");
};

//Client
void ClientLandingRequest(Airplane* airplane){
    airplane->requestLanding();
}
//Client
void ClientTakeoffRequest(Airplane* airplane){
    airplane->requestTakeoff();
}

int main(){
    ControlTower* controlTower = new AirportControlTower();
    Airplane* airplane1 = new CommercialAirplane(controlTower);
    Airplane* airplane2 = new MilitaryAirplane(controlTower);

    ClientLandingRequest(airplane1);
    ClientTakeoffRequest(airplane2);

    delete controlTower;
    delete airplane1;
    delete airplane2;
    return 0;
}
```

10.4.4 备忘录方法

备忘录方法是一种行为设计模式，用于在不违反封装的情况下捕获和恢复对象的状态，允许程序员将对象的状态保存后恢复到以前的状态，从而撤销或回滚对对象所做的更改。

备忘录方法通常包括以下几个组成部分：

（1）Originator：通常是一个组件，负责创建和管理对象的状态。Originator 包含设置和获取对象状态的方法，可以创建 Memento 对象来存储状态。Originator 直接与 Memento 对象通信，以创建该对象的状态快照，并从状态快照中恢复该对象的状态。

（2）Memento：是一个对象，用于存储 Originator 在特定时间点的状态。Memento 对象只提供一种检索状态的方法，不允许直接修改状态，这确保状态可以保持不变。

（3）CareTaker：负责跟踪 Memento 对象，它不知道存储在 Memento 对象中的状态详细信息，但可以向 Originator 请求 Memento 对象以保存或恢复对象的状态。

（4）Client：通常表示应用程序或系统的一部分，它与 Originator 和 CareTaker 交互以实现特定功能。Client 通过 CareTaker 发起保存或恢复 Originator 对象状态的请求。

图 10.4-4 所示为备忘录方法的 UML 类图示例，图中，Memento 是接口类，在 C++中，Memento 是抽象基类，它定义了一个接口函数 getSavedContent()，该接口函数是纯虚函数。类 TextEditorMemento 是一个 ConcreteMemento（具体备忘录类），由 ConcreteMemento 实现接口函数 getSavedContent()的细节。类 TextEditor 是一个 ConcreteOriginator（具体原发器类），由

ConcreteOriginator 实现类 Originator 中的接口函数 createMemento()和 restoreFromMemento()。类 TextEditor 的主要功能是文本编辑。类 CareTaker 负责保存所有文本编辑器的快照，这是通过函数 Backup()来实现保存的。类 CareTaker 向 Client 提供 Undo()调用函数。为了简单起见，Client 的代码只有一个函数 clientCall()，在该函数中创建了类 CareTaker 的对象（careTaker）和类 TextEditor 的对象（originator），并模拟备份（Backup）操作和撤销（Undo）操作。

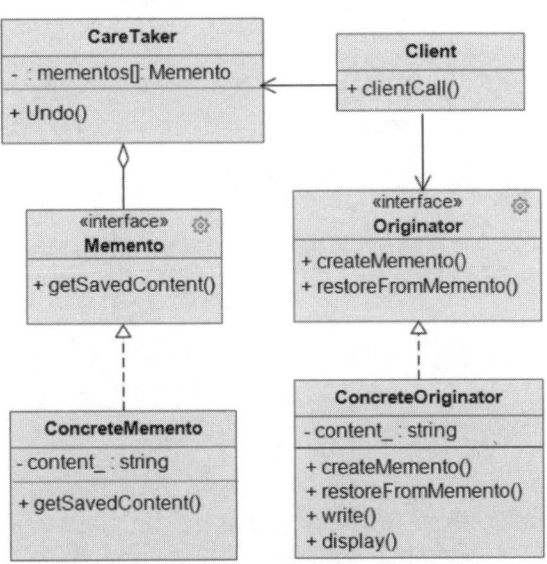

图 10.4-4　备忘录方法的 UML 类图示例

【例 10-16】

```
#include<iostream>
#include<vector>
#include<string>

class Memento{
public:
    virtual ~Memento() {}
    virtual std::string getSavedContent() =0;
};

//ConcreteMemento
class TextEditorMemento: public Memento{
private:
    std::string content_;
public:
    TextEditorMemento(std::string content){ content_ = content;}
    std::string getSavedContent()override{ return content_;}
};

//Originator
class TextEditor{
private:
```

```cpp
    std::string content_;
public:
    TextEditor(std::string content) { content_ = content;}
    void write(std::string text) { content_ += text;}
    string getContent() { return content_;}
    Memento* createMemento() {
        return new TextEditorMemento(content_);
    }
    void restoreFromMemento(Memento* memento) {
        content_ = memento->getSavedContent();
    }
    void display() const{std::cout << content_;}
};

//CareTaker
class CareTaker{
private:
    std::vector<Memento*> mementos_;
    TextEditor *originator_;
public:
    CareTaker(TextEditor *originator), originator_(originator){}
    ~Caretaker() { for (auto m : mementos_) delete m;}
    void Backup() {
        mementos_.push_back(originator_->createMemento());
    }
    Memento* getlastMemento() {
        if (mementos_.size()>0) {
            mementos_.pop_back();
            Memento *m = mementos_.back();
            return  m;
        }
        return nullptr;
    }
    void Undo(){
        Memento *m = getlastMemento();
        if(m){
            originator_-> restoreFromMemento(m);
            originator_->display();
        }
    }
    void ShowHistoryContent() const {
        for (Memento *m : mementos_) {
            std::cout << m->getSavedContent()<< "\n";
        }
    }
};

void ClientCode(){
```

```cpp
        TextEditor* originator = new TextEditor("Hello world.");
        Caretaker * caretaker = new Caretaker(originator);
        caretaker->Backup();
        originator->write("first");
        caretaker->Backup();
        originator->write("second");
        caretaker->Backup();
        originator->write("third");
        std::cout << "\n";
        caretaker->ShowHistoryContent();
        caretaker->Undo();
        caretaker->Undo();
        delete originator;
        delete caretaker;
}

int main(){
        ClientCode();
        return 0;
}
```

10.4.5 观察者方法

观察者方法是一种行为设计模式，它定义了对象之间的一对多依赖关系，以便当一个主题（Subject）更改状态时，其所有的依赖对象（Observer）都会自动收到通知和更新。观察者方法具有以下的特点：

（1）观察者方法定义了一组对象基于某个主题（Subject）状态变化进行交互的机制。Observer 的行为是由 Subject 状态的变化触发的。

（2）观察者方法封装了依赖对象（Observer）的行为，并允许 Subject 与 Observer 之间完全隔离。这种隔离实现了模块化并提高了设计的可维护性。

（3）观察者方法实现了 Subject 和 Observer 之间的松耦合。Subject 不需要知道 Observer 的具体类别，可以在不影响 Subject 的情况下添加或删除 Observer。

（4）观察者方法中的主要机制是在 Subject 的状态发生变化时通知 Observer，这种通知机制促进了多个对象的动态协调行为，以响应 Subject 的变化。

在现实生活中，如果我们订阅了报纸或杂志，则无须查看下一期报纸或杂志是否有货，出版商会将新一期报纸或杂志直接邮寄给我们。出版商维护着一个订阅者列表，并知道这些订阅者对哪些报纸或杂志感兴趣。当订阅者不希望出版商再向他们邮寄新的报纸或杂志时，他们可以随时退出订阅者列表。

观察者方法通常包括以下组成部分：

（1）Subject：提供了动态注册和注销 Observer 的方法，并定义了一种通知 Observer 其状态变化的方法。

（2）Observer：定义了 ConcreteObserver 从 Subject 接收更新的通用或一致的接口函数，ConcreteObserver 实现了这个接口函数，并对 Subject 的状态变化做出反应。

（3）ConcreteSubject：是 Subject 的特定实现，并维护一个订阅者列表，订阅者列表保存了

Observer 想要跟踪的实际状态或数据。当 Subject 的状态发生变化时，会通知 Observer。

（4）ConcreteObserver：实现 Observer 的接口函数，注册了一个 ConcreteSubject，并在收到状态变化的通知时做出反应。当 Subject 的状态发生变化时，将调用 ConcreteObserver 的 Update()方法，使 Observer 能够采取适当的操作。

图 10.4-5 所示为观察者方法的 UML 类图示例。图中，TVObserver 和 MobilephoneObserver 是 Observer 的派生类。智能手机上的应用程序是一个 ConcreteObserver，TV 上的应用程序是另一个 ConcreteObserver。这两个 ConcreteObserver 都各自实现了接口函数 Update()，同时它们还实现了自己的私有函数。接口类 Subject 有三个接口函数，即 Add()、Remove()和 Notify()，这三个接口函数都是在派生类 SportsSubject 中具体实现的。派生类 SportsSubject 中还有订阅者列表。

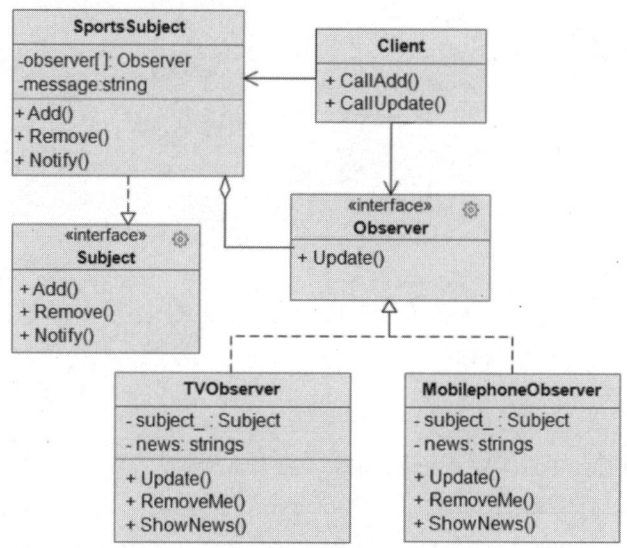

图 10.4-5 观察者方法的 UML 类图示例

【例 10-17】

```
#include <iostream>
#include <vector>
#include <string>

class Observer {
public:
    virtual void Update(const std::string &message)= 0;
};

class Subject{
public:
    virtual void Add(Observer *observer) = 0;
    virtual void Remove(Observer *observer) = 0;
    virtual void Notify() = 0;
};

//ConcreteSubject
class SportsSubject : public Subject {
```

```cpp
public:
    virtual ~SportsSubject(){
        std::cout << "Goodbye, I was the Subject.\n";
    }
    void Add(Observer *observer) override {
        observers_.push_back(observer);
    }

    void Remove(Observer *observer) override {
        std::vector<Observer*>::iterator iter = observers_.begin();
        for (; iter != observers_.end(); ++iter;) {
            if(*iter == observer)
            observers_.erase(it);
        }
    }

    void Notify() override {
        std::vector<Observer*>::iterator iter = observers_.begin();
        while (iter != observers_.end()){
            (*iter)->Update(message_);
            ++iter;
        }
    }

    void PublishMessage(std::string message){
        this->message_ = message;
        Notify();
    }

    void getObservers() {
        return observer_.size();
    }
private:
    std::vector<Observer *> observers_;
    std::string message_;
};

//ConcreteObserver
class TVObserver : public Observer {
public:
    TVObserver(Subject *subject) : subject_(subject){
        subject_->Add(this);
        std::cout << "Hi, I'm the TV Observer."<<endl;
    }

    void Update(const std::string & message) override{
        message_ = message;
        ShowMessage();
    }
```

```cpp
        void RemoveMe() {
            ubject_->Remove(this);
        }

        void ShowMessage() {
            std::cout <<"a TV message:"<< message_ << endl ;
        }
    private:
        std::string message_;
        Subject *subject_;
};

//ConcreteObserver
class MobilePhoneObserver : public Observer {
public:
    MobilePhoneObserver(Subject* subject) : subject_(subject){
        subject_->Add(this);
        std::cout << "Hi, I'm the Mobile Observer."" << endl;
    }
    void Update(const std::string &message) override {
        message_ = message;
        ShowMessage();
    }
    void RemoveMe(){
        subject_->Remove(this);
    }
    void ShowMessage() {
        std::cout << "a mobile message:" << message_ << endl;
    }
private:
    std::string message_;
    Subject* subject_;
};

void ClientCode(){
    Subject *subject = new SportsSubject;
    Observer *observer1 = new TVObserver(subject);
    Observer *observer2 = new MobilePhoneObserver(subject);
    subject->Add(observer1);
    subject->Add(observer2);
    subject->PublishMessage("Sports News");
    subject->PublishMessage("The weather is very nice!");

    observer2->RemoveMe();
    observer1->RemoveMe();
    delete observer2;
    delete observer1;
    delete subject;
```

```
}
int main() {
    ClientCode();
    return 0;
}
```

10.4.6　状态方法

状态方法是一种行为设计模式，允许对象在其内部状态变化时更改行为，这是通过将对象的行为封装在不同的状态对象中来实现的，并且对象本身可以根据其当前状态在不同的状态对象之间动态切换。

假设我们在自动饮料售货机上买一瓶可乐，自动饮料售货机开始处于准备（Ready）状态，当我们投币或扫码后，自动饮料售货机进入一个新的状态，让我们选择饮料，不妨称为选择（Select）状态；当我们选择可乐后，自动饮料售货机进入下一状态送出饮品，不妨称为输出（Output）状态；当我们拿起饮料后，自动饮料售货机进入下一状态找零钱（Changes）状态；当我们拿走零钱后，自动饮料售货机回到准备状态。

状态方法通常包括以下组成部分：

（1）Context：包含 State 对象，State 对象的行为会根据其内部状态变化而更改。Context 维护着对当前 State 对象的引用，该对象表示 Context 的当前状态。Context 为 Client 提供了一个接口函数，以便与之进行交互，并通过这个接口函数表示 State 对象的行为。

（2）State：State 是接口基类，它定义了所有 ConcreteState 的公共接口，该公共接口通常被声明为一个函数，表示 Context 在特定状态下的行为。

（3）ConcreteState：类 State 的派生类，通常有多个，每个 ConcreteState 都以不同的方式实现了类 State 的接口函数。每个 ConcreteState 都封装了与上下文特定状态关联的行为，这些 ConcreteStates 定义了 Context 在不同状态下的行为方式。

状态方法的 UML 类图示例如图 10.4-6 所示。

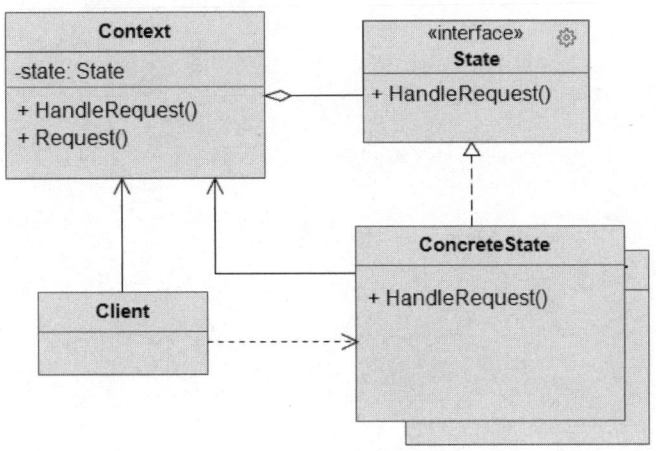

图 10.4-6　状态方法的 UML 类图示例

例 10-18 给出了采用状态方法的自动饮料售货机原型系统的 C++ 代码。

【例 10-18】

```cpp
#include <iostream>
#include <string>
class VendingMachineState;

class VendingMachineContext{
public:
    void setState(VendingMachineState* state) {
        if(state_) delete state_;
        state_ = state;
    }
    void request(){
        state_->handleRequest();
    }
private:
    VendingMachineState* state_;
};

class VendingMachineState {
public:
    void handleRequest() = 0 ;
};

class VendingMachineReadyState : public VendingMachineState {
public:
    void handleRequest() {
        std::cout << "Ready: Please insert coin." << endl;
    }
};

class VendingMachineSelectState : public VendingMachineState {
public:
    void handleRequest(){
        std::cout << "select: Please select a drink." << endl;
    }
};

class VendingMachineOutputState : public VendingMachineState {
public:
    void handleRequest() {
        std::cout << "Output: Please take your drink." << endl;
    }
};
class VendingMachineChangesState : public VendingMachineState {
public:
    void handleRequest() {
        std::cout <<"Changes: Please take your changes." << endl;
    }
```

```cpp
};
void ClientCode() {
    Context *context = new VendingMachineContext(new VendingMachineReadyState);
    context->request();
    context->setState( new VendingMachineSelectState );
    context->request();
    context->setState( new VendingMachineOutputState );
    context->request();
    context->setState( new VendingMachineChangesState )
    context->request();
    context->setState(nullptr);

    delete context;
}

int main() {
    ClientCode();
    return 0;
}
```

10.4.7 策略方法

策略方法是一种行为设计模式，该方法定义并封装了一系列算法，并使这些算法可以互换，从而允许客户端在不更改代码结构的情况下动态切换算法。

策略方法允许在运行时选择对象的行为，它也是"四人组"（Gang of Four，GoF）设计模式之一，广泛用于面向对象编程。简单来说，通过策略方法，程序员可以定义并封装一系列算法，并互换这些算法。策略方法允许算法独立于使用它的客户端。

策略方法通常包括以下组成部分：

（1）Context：是一个类或对象，它包含对 Strategy 对象的引用并将任务委托给 Strategy 对象，充当 Client 和 Strategy 对象之间的接口，提供统一的方式来执行任务，而无须了解完成任务的细节。

（2）Strategy：是一个接口类，它定义了所有 ConcreteStrategy 必须实现的一组方法。接口类 Strategy 充当一个契约，确保所有的 Strategy 对象都遵循相同的规则集，并且可以由 Context 互换使用。通过定义一个通用接口函数，Strategy 允许在 Context 和 ConcreteStrategy 之间解耦，从而促进设计的灵活性和模块化。

（3）ConcreteStrategy：是接口类 Strategy 接口函数的各种实现。每个 ConcreteStrategy 都提供特定的算法或行为，用于执行接口类 Strategy 接口函数定义的任务。ConcreteStrategy 封装了各自算法的细节，并提供了执行任务的方法，它们是可以互换的，可以由 Client 根据任务要求进行选择和配置。

（4）Client：Client 负责选择和配置适当的 ConcreteStrategy，并将其提供给 Context。

图 10.4-7 所示为策略方法的 UML 类图示例。Strategy 是接口类，定义了接口函数 Execute()。该接口函数是在 ConcreteStrategy 中实现的。ConcreteStrategy 一般有多个，每一个都是由接口类 Strategy 派生出来的，每一个的接口函数 Execute()的实现各不相同。Context 中有 Strategy 对象，Client 通过 Strategy 对象来调用接口函数 Execute()。

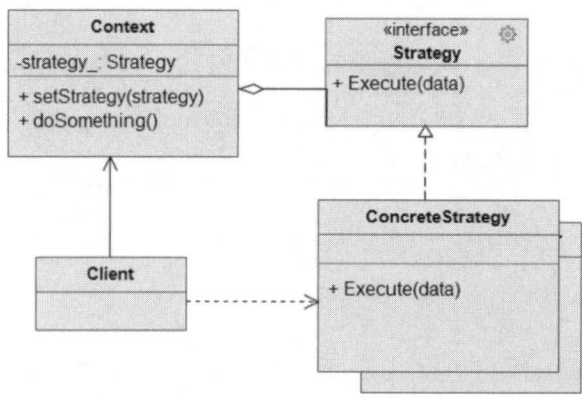

图 10.4-7 策略方法的 UML 类图示例

例 10-19 给出了一个采用策略方法的运输方案原型系统的 C++语言代码。DeliveryStrategy 是接口类 Strategy，定义的接口函数为 Deliver()，该函数是一个纯虚函数。三个 ConcreteStrategy 分别是 TruckDeliveryStrategy、ShipDeliveryStrategy 和 AirplaneDeliveryStrategy，它们以不同的方式实现了接口 Deliver()。DeliveryContext 对应类 Context，定义了 Strategy 对象和向 Client 提供的函数。

【例 10-19】

```cpp
#include <iostream>
#include <string>

class DeliveryStrategy {
public:
    virtual void Deliver() = 0;
};

class TruckDeliveryStrategy :public DeliveryStrategy {
public:
    void Deliver(){
        std:: cout<< "Deliver by truck" << endl;
    }
};

class ShipDeliveryStrategy : public DeliveryStrategy {
public:
    void Deliver(){
        std:: cout<< "Deliver by ship" << endl;
    }
};

class AirplaneDeliveryStrategy : public DeliveryStrategy {
public:
    void Deliver(){
        std:: cout<< "Deliver by airplane" << endl;
    }
};
```

```cpp
class DeliveryContext{
private:
    DeliveryStrategy* deliveryStrategy_;

public:
    DeliveryContext(DeliveryStrategy* ds) {
        deliveryStrategy_ = ds;
    }

    virtual ~DeliveryContext(){
        if(deliveryStrategy_)
        delete deliveryStrategy_;
    }

    void setDeliveryStrategy(DeliveryStrategy* ds) {
        if(deliveryStrategy_)
        delete deliveryStrategy_;
        deliveryStrategy_ = ds;
    }

    void DoDelivery() {
        if(deliveryStrategy_)
        deliveryStrategy_->Deliver();
    }
};

void clientCode(){
    DeliveryContext context(new TruckDeliveryStrategy);
    context.DoDelivery();
    context.setDeliverystrategy(new ShipDeliveryStrategy);
    context.DoDelivery();
    context.setDeliverystrategy(new AirplaneDeliveryStrategy);
    context.DoDelivery();
}

int main(){
    clientCode();
    return 0;
}
```

10.4.8　模板方法

模板方法是一种行为设计模式，它在一个超类中定义了整个算法框架，允许子类在不更改超类算法框架的情况下覆盖算法的特定步骤，通过将通用算法封装在超类中来实现代码重用，同时允许子类为某些步骤提供具体实现，从而实现自定义和灵活性。

假设我们需要开发一个数据分析软件，需要面对文本数据、音频数据、图像数据。我们设计一个通用的数据分析模板算法框架，主要包括数据加载、数据去噪、特征提取、分析算法和结果输出，并对外提供统一的接口。算法框架不能变动，针对每一种数据，相应的函数实现是不同的。

模板方法通常包括以下几个组成部分：

（1）AbstractClass：这是定义 TemplateMethod 的超类，它为算法提供了一个框架，其中定义了某些步骤，但其他步骤是抽象的或者是子类可以覆盖的钩子，还可能包括所有子类通用的具体方法，并在 TemplateMethod 中使用。

（2）TemplateMethod：这是 AbstractClass 中的方法，它按特定顺序调用各种步骤来定义整个算法框架。它通常被声明为 final 类型，以防止子类更改算法框架。TemplateMethod 通常由一系列方法调用（抽象或具体）组成，这些方法调用构成了算法的步骤。

（3）Hook：是在 AbstractClass 中声明但未实现的方法，由子类 ConcreteSubClass 来实现细节。子类必须为这些方法提供具体的实现才能完成算法。

（4）ConcreteSubClass：这些子类扩展了 AbstractClass，并为 AbstractClass 中定义的抽象方法提供了具体实现。每个子类都可以覆盖算法的某些步骤，在不更改算法框架的情况下自定义行为。

模板方法的 UML 类图示例如图 10.4-8 所示。

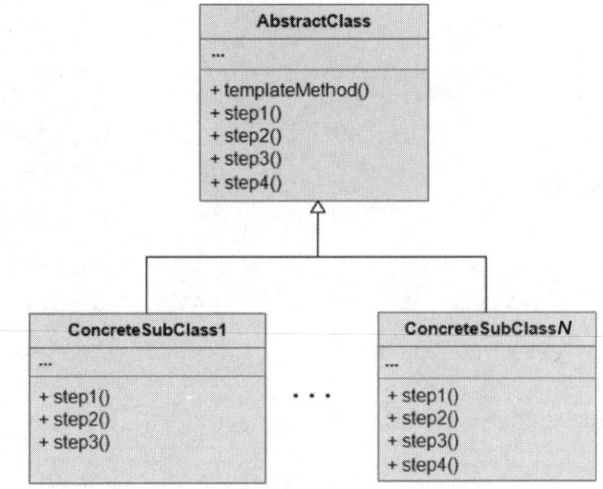

图 10.4-8　模板方法的 UML 类图示例

例 10-20 中的 C++代码是基于果汁（Fruit Juice）制作过程展开的。类 FruitJuiceMaker 是 AbstractClass，其中定义了 makeFruitJuice 算法框架，即 TemplateMethod。makeFruitJuice 算法框架的内部顺序结构不能改动。AppleJuiceMaker 和 OrangeJuiceMaker 是 ConcreteSubClass。FruitSorting()和 FruitWashing()等是 Hook 函数，它们将在 AppleJuiceMaker 和 OrangeJuiceMaker 中实现。

【例 10-20】

```cpp
#include <iostream>
#include <string>

class FruitJuiceMaker {
    //使用模板方法（TemplateMethod）定义整个过程
public:
    void makeFruitJuice() {
```

```cpp
            FruitSorting();
            FruitWashing();
            JuiceExtraction();
            Clarifying();
            Pasteurisation();
            Packing();
        }

        virtual void FruitSorting() = 0;
        virtual void FruitWashing() = 0;
        virtual void JuiceExtraction() = 0;
        virtual void Clarifying() = 0;
        virtual void Pasteurisation() = 0;
        virtual void Packing() = 0;
};

class AppleJuiceMaker extends FruitJuiceMaker {
    //实现抽象的方法
public:
        void FruitSorting(){
            std::cout << "pick appropriate apples" << endl;
        }
        void FruitWashing(){
            std::cout << "wash the picked apples" << endl;
        }
        void JuiceExtraction(){
            std::cout << "extract juice from washed apples" << endl;
        }
        void Clarifying(){
            std::cout << "clearfy the apple juice" << endl;
        }
        void Pasteurisation(){
            std::cout << "do apple juice pasteurisation" << endl;
        }
        void Packing(){
            std::cout << "do apple juice packing" << endl;
        }
};

class OrangeJuiceMaker extends FruitJuiceMaker {
    //实现抽象的方法
public:
        void FruitSorting(){
            std::cout << "pick appropriate oranges" << endl;
        }
        void FruitWashing(){
            std::cout << "wash the picked oranges" << endl;
        }
        void JuiceExtraction(){
```

```cpp
            std::cout << "extract juice from washed oranges" << endl;
        }
        void Clarifying(){
            std::cout << "clearfy the orange juice" << endl;
        }
        void Pasteurisation(){
            std::cout << "do orange juice pasteurisation" << endl;
        }
        void Packing(){
            std::cout << "do orange juice packing" << endl;
        }
};

void ClientCode(FruitJuiceMaker *juiceMaker_){
    juiceMaker_->makeFruitJuice();
}

int main(){
    std::cout << "make apple juice.\n";
    FruitJuiceMaker * appleJuice = new AppleJuiceMaker;
    ClientCode(appleJuice);

    std::cout << "make orange juice\n";
    FruitJuiceMaker * orangeJuice = new OrangeJuiceMaker;
    ClientCode(orangeJuice);

    delete appleJuice;
    delete orangeJuice;
    return 0;
}
```

10.4.9　命令方法

命令方法是一种行为设计模式，它可将请求转换为独立对象，允许对具有不同请求的 Client 进行参数化、对请求进行排队以及支持不可撤销的操作。

命令方法在软件产品本身有很多的应用。例如文本编辑器，它有 Copy、Cut、Paste 等命令；再如一些 GUI 控件，它们可能有 Click、DoubleClick、KeyDown、KeyUp 等命令。这些都可以使用命令方法进行设计。

命令方法通常包括以下组成部分：

（1）Command：接口类 Command 是所有的命令类都遵循的"规则手册"，它声明了一个通用方法 execute()，确保每个 ConcreteCommand（具体命令）都知道如何执行其特定的操作。接口类 Command 为所有的命令设定了标准，使 Invoker（调用程序）更容易管理和执行各种操作，而无须了解每个命令的详细信息。

（2）ConcreteCommand：类 ConcreteCommand 是特定的命令，如打开电视或调整音量。每个 ConcreteCommand 都封装了特定操作的详细信息，这些类充当 Invoker 可以触发的可执行指令，而无须担心每个命令是如何完成其任务细节的。

(3) Invoker：是负责启动命令执行的调用程序，它包含对命令的引用，但不深入研究每个命令的工作细节。Invoker 就像一个按钮，当按下时，事情就会发生。Invoker 的作用是协调和执行命令，而无须考虑单个操作的复杂性。

(4) Receiver：是知道如何执行与命令相关的实际操作的对象。Receiver 了解命令中提到的特定任务并确切地知道如何执行该操作。Receiver-Command 关系将职责分开，从而可以轻松添加新 Receiver 或 ConcreteCommand，而不会弄乱现有功能。

命令方法的 UML 类图示例如图 10.4-9 所示。

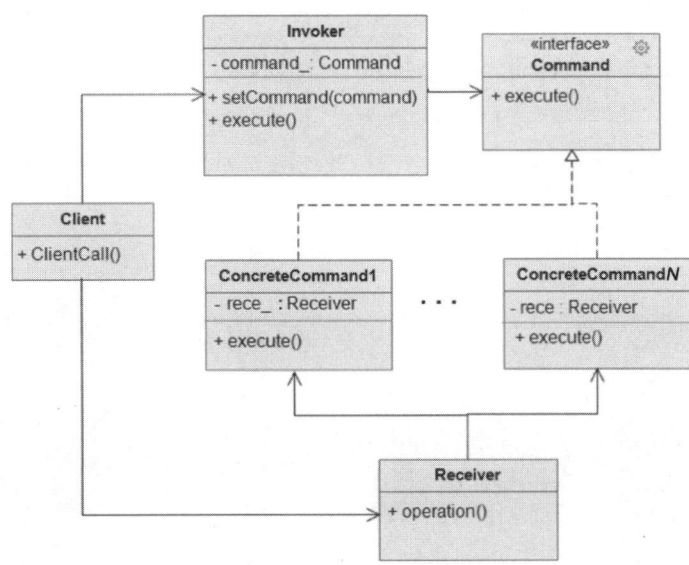

图 10.4-9　命令方法的 UML 类图示例

例 10-21 是采用命令方法的文本编辑示例的 C++代码，它覆盖了 Copy、Cut 和 Paste 三个 ConcreteCommand，TextEditor 和 Clipboard 两个 ConcreteReceiver，以及 CopyCommand、CutCommand 和 PasteCommand 三个 ConcreteCommand。

【例 10-21】

```
#include <iostream>
#include <string>

class Command {
public:
    virtual ~Command() {}
    virtual void Execute() = 0;
};

//类 Editor，相当于 Receiver
class Editor{
public:
    virtual ~Editor() {}
    virtual std::string Copy(){return "";}
    virtual std::string Cut(){return "";}
    virtual std::string Paste(){return "";}
    virtual void Copy(std::string copyContent){}
```

```cpp
        virtual void Cut(std::string cutContent){}
        virtual void Paste(std::string pasteContent){}
};

//类 CopyCommand,相当于 ConcreteCommand
class CopyCommand : public Command{
private:
    Editor* textEditor_;
    Editor* clipboard_;
public:
    explicit CopyCommand(Editor* textEditor, Editor* clipboard){
        textEditor_ = textEditor;
        clipboard_  = clipboard;
    }
    void Execute(){
        std::string temp;
        temp = textEditor_->Copy();
        clipboard_->Copy(temp);
    }
};

//类 CutCommand,相当于 ConcreteCommand
class CutCommand : public Command {
private:
    Editor* textEditor_;
    Editor* clipboard_;
public:
    explicit CutCommand(Editor* textEditor, Editor* clipboard){
        textEditor_ = textEditor;
        clipboard_  = clipboard;
    }

    void Execute(){
        std::string temp;
        temp = textEditor_->Cut();
        clipboard_->Cut(temp);
    }
};

//类 PasteCommand,相当于 ConcreteCommand
class PasteCommand : public Command {
private:
    Editor* textEditor_;
    Editor* clipboard_;
public:
    explicit PasteCommand(Editor*textEditor, Editor*clipboard){
        textEditor_ = textEditor;
        clipboard_  = clipboard;
    }
```

```cpp
    void Execute(){
        std::string temp;
        Temp = clipboard_->Paste();
        textEditor_->Paste(Temp);
    }
};

//类 TextEditor，相当于 ConcreteReceiver
class TextEditor : public Editor{
private:
    std::string content_;
public:
    void Add(const std::string str) {content_ += str;}

    std::string Copy(){
        return content_;
    }

    std::string Cut(){
        std::string clipboard = content_;
        content_ = "";
        return clipboard;
    }
    void paste(std::string clipboard){
        content_ += clipboard;
    }
};

//类 Clipboard，相当于 ConcreteReceiver
class Clipboard : public Editor{
private:
    std::string clipboard_;
public:
    void Copy(std::string copyContent) {
        clipboard_ = copyContent;
    }

    void Cut(std::string cutContent) {
        clipboard_ = cutContent;
    }

    std::string Paste(){
        return clipboard_;
    }
};

class Invoker {
private:
```

```cpp
        Command *on_start_;
        Command *on_finish_;
public:
    ~Invoker(){
        delete on_start_;
        delete on_finish_;
    }

    void SetOnStart(Command *command) {
        this->on_start_ = command;
    }

    void SetOnFinish(Command *command) {
        this->on_finish_ = command;
    }

    void DoTask() {
        if (this->on_start_) {
            this->on_start_->Execute();
        }
        if (this->on_finish_) {
            this->on_finish_->Execute();
        }
    }
};

//Client 的代码放在 main()里
int main(){
    Editor* textEditor = new TextEditor();
    Editor* clipboard =   new Clipboard();
    textEditor.Add("Command Method Design Pattern");

    Command* copyCommand = new CopyCommand(textEditor, clipboard);
    Command* pasteCommand = new PasteCommand(textEditor, clipboard);

    Invoker* invoker = new Invoker();
    invoker->setOnStart(copyCommand);
    invoker->setOnStart(pasteCommand);
    invoker->DoTask();

    delete invoker;
    delete clipboard;
    delete textEditor;
    return 0;
}
```

10.4.10 访客方法

访客方法是行为设计模式之一。当我们必须对一组类型相似的对象进行操作时，可采用访客方法将操作逻辑从对象剥离并移到另一个类中。

访客方法通常有两个函数：一个是名为 visit() 的函数，由访客（Visitor）实现，可以由数据结构中的每个元素（即对象）调用；另一个是接受访客的 accept() 函数。

访客方法通常包含以下几个组成部分：

（1）Visitor：这是一个接口类，为所有类型访客的可访问类声明访问函数。

（2）ConcreteVisitor：对于每种类型访客，必须实现所有在接口类 Visitor 中声明的访问方法，每个 ConcreteVisitor 负责不同的操作。

（3）Element：这是一个声明接受操作的接口类，是使 ConcreteVisitor 能够访问对象的入口点。

（4）ConcreteElement：这些类用于实现接口类 Element 并定义 accept()，ConcreteVisitor 使用 accept() 操作将 ConcreteElement 传递给此对象。

（5）Client：是访客方法中的使用者，它有权访问数据结构对象，并可以指示它们接受访客以执行适当的处理。

访客方法的 UML 类图示例如图 10.4-10 所示。

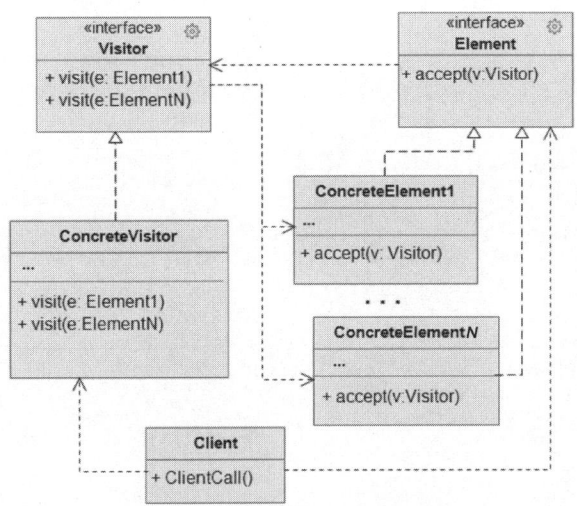

图 10.4-10　访客方法的 UML 类图示例

例 10-22 给出了采用访客方法的菜市场场景的 C++ 代码，其中，Market 是接口类，我们在其中定义了两个元素，即 Tomato 和 Potato，这两个元素要实现接口类 Market 中的 accept()。接口类 Market 有两种 ConcreteVisitor，一个是 VisitBuyer，另一个是 VisitSeller。VisitBuyer 和 VisitSeller 都会访问两个元素 Tomato 和 Potato。例 10-22 模拟了 Tomato 和 Potato 的价格和交易量。

【例 10-22】

```
#include <iostream>
using namespace std;

class Market{
public:
    virtual void accept(class Visitor *, int kilos) = 0;
};

class Tomato : public Market{
public:
```

```cpp
        Tomato(int price) : price_(price){}

        void accept(Visitor * v, int kilos){
            v->visit(this, kilos);
        }

        void buy(int kilos){
            cout << "Tomato::buy" << kilos * price_ << endl;
        }

        void sell(int kilos){
            cout << "Tomato::sell" << kilos * price_ << endl;
        }

private:
    int price_;
};

class Potato : public Market{
public:
        Potato(int price) : price_(price){}
        void accept(Visitor * v, int kilos){
            v->visit(this, kilos);
        }

        void buy(int kilos){
            cout << " Potato::buy : " << kilos * price_ << endl;
        }

        void sell(int kilos){
            cout << " Potato::sell: " << kilos* price_ << endl;
        }

private:
    int price_;
};

class Visitor{
public:
    virtual void visit(Tomato *, int kilos) = 0;
    virtual void visit(Potato *, int kilos) = 0;
};

class BuyVisitor : public Visitor{
public:
    BuyVisitor(){
        tomato_ = 0;
        potato_ = 0;
    }
```

```cpp
        void visit(Tomato *t, int kilos){
            tomato_ = kilos;
            t->buy(tomato_);
            cout << "tomato buyed:" << tomato_ << endl;
        }
        void visit(Potato *p, int kilos){
            potato_ = kilos;
            p->buy(potato_);
            cout << "Potato buyed:" << potato_ << '\n';
        }

    private:
        int tomato_;
        int potato_;
};

class SellVisitor : public Visitor{
    public:
        SellVisitor(){
            tomato_stock_ =20;
            potato_stock_ = 20;
        }

        void visit(Tomato *t, int kilos) {
            tomato_stock_ -= kilos;
            t->sell(kilos);
            cout << "tomato selled:" << kilos << endl;
        }

        void visit(Potato *p, int kilos){
            potato_stock_ -= kilos;
            p->sell(kilos);
            cout << "potato selled:" << kilos << endl;
        }

        int tomato_stock_;
        int potato_stock_;
};

int main(){
    Market *vegetables[2] = { new Tomato(6), new Potato(4)};

    BuyVisitor buyer;
    SellVisitor seller;

    //市场上的 Tomato 交易了 5 kg
    vegetables[0]->accept(&buyer, 5);
    vegetables[0]->accept(&seller, 5);
```

```cpp
        //市场上的 potato 交易了 8 kg
        vegetables[1]->accept(&buyer, 8);
        vegetables[1]->accept(&seller, 8);

        delete [] vegetables;
        return 0;
}
```

10.5 本章小结

　　C++的设计模式主要包括创建设计模式、结构设计模式和行为设计模式。每种设计模式又包含一系列设计方法，本章共计介绍了 22 种设计方法。

　　本章对每种设计方法的用途、组成、UML 类图做了详细介绍，并给出了每种设计方法的示例。需要注意的是：每种设计方法示例中的 C++代码是示例性的，很多接口函数的功能并没有实现，示例重点展示的是每种设计方法的 C++实现。

　　C++中的设计模式可帮助开发人员创建可维护、更灵活且易于理解的代码。这些设计模式囊括了软件架构师和开发人员的专业知识和丰富经验，可使新程序员更容易地遵循既定的最佳实践。

　　为了更好地理解设计模式，需要读者熟悉 UML 的相关知识，特别是 UML 类图。本章在开始处介绍了 UML 类图的基本知识，这些基础知识是软件架构师必备的，感兴趣的读者可以进一步阅读 UML 的相关文献。

　　对于 C++高级程序员和资深软件工程师来说，熟悉并掌握设计模式并把它们应用到软件开发中，会大大增加代码的灵活性、可重用性和可维护性，可使代码在整体上更加艺术化。

参考文献

[1] LIPPMAN S, LAJOIE J, MOO B. C++ primer[M]. 5th ed. Boston, MA: Addison-Wesley Professional, 2012.

[2] STROUSTRUP B. Programming: principles and practice using C++[M]. 3rd ed. Boston, MA: Addison-Wesley, 2024.

[3] MEYERS S. Effective modern C++: 42 specific ways to improve your use of C++11 and C++14[M]. Sebastopol, CA: O'Reilly Media, 2014.

[4] GREGOIRE M. Professional C++[M]. Hoboken, NJ: Wrox, 2021.

[5] BANCILA M. Modern C++ programming cookbook[M]. Birmingham: Packt Publishing, 2020.

[6] GRIMES R, BANCILA M. Modern C++: efficient and scalable application development[M]. Birmingham: Packt Publishing, 2018.

[7] WILLIAMS A. C++ concurrency in action[M]. Greenwich, CT: Manning Publications, 2012.

[8] GAMMA E, HELM R, JOHNSON R, et al. Design patterns: elements of reusable object-oriented software[M]. Boston, MA: Addison-Wesley Professional, 1994.

[9] ZEROMSKI A. 15 most used design patterns: a design patterns cheat book[M]. Boston, MA: Addison-Wesley Professional, 2023.

[10] FREEMAN E, BATES B, SIERRA K, et al. Head first design patterns[M]. Boston, MA: Addison-Wesley Professional, 2004.

[11] SANDERS J, KANDROT E. CUDA by example: an introduction to general-purpose GPU programming[M]. Boston, MA: Addison-Wesley Professional, 2010.

[12] COOK S. CUDA programming: a developer's guide to parallel computing with GPUs[M]. Burlington, MA: Morgan Kaufmann, 2012.

[13] MUNSHI A. OpenCL programming guide[M]. Boston, MA: Addison-Wesley Professional, 2011.

[14] BANGER R, BHATTACHARYYA B. OpenCL programming by example[M]. Birmingham: Packt Publishing, 2013.